Making a Difference

TRANSDISCIPLINARY STUDIES
Volume 1

Scope
Transdisciplinary Studies is an internationally oriented book series created to generate new theories and practices to extricate transdisciplinary learning and research from the confining discourses of traditional disciplinarities. Within transdisciplinary domains, this series publishes empirically grounded, theoretically sound work that seeks to identify and solve global problems that conventional disciplinary perspectives cannot capture. Transdisciplinary Studies seeks to accentuate those aspects of scholarly research which cut across today's learned disciplines in an effort to define the new axiologies and forms of praxis that are transforming contemporary learning. This series intends to promote a new appreciation for transdisciplinary research to audiences that are seeking ways of understanding complex, global problems that many now realize disciplinary perspectives cannot fully address. Teachers, scholars, policy makers, educators and researchers working to address issues in technology studies, education, public finance, discourse studies, professional ethics, political analysis, learning, ecological systems, modern medicine, and other fields clearly are ready to begin investing in transdisciplinary models of research. It is those many different audiences in these diverse fields that we hope to reach, not merely with topical research, but also through considering new epistemic and ontological foundations for of transdisciplinary research. We hope this series will exemplify the global transformations of education and learning across the disciplines for years to come.

Making a difference
Science, action and integrated environmental research

Lorrae van Kerkhoff

The Australian National University, Canberra

SENSE PUBLISHERS
ROTTERDAM / TAIPEI

A C.I.P. record for this book is available from the Library of Congress.

ISBN 978-90-8790-391-6 (paperback)
ISBN 978-90-8790-392-3 (hardback)
ISBN 978-90-8790-393-0 (e-book)

Published by: Sense Publishers,
P.O. Box 21858, 3001 AW
Rotterdam, The Netherlands

Printed on acid-free paper

This monograph is dedicated to my grandmother, Hazel Sheedy, who passed away shortly before its completion.

TABLE OF CONTENTS

PREFACE

It's been a long question and answer session, and the convener calls you out for the last question before the reception. You hesitate between the rich and complex comment you want to make and the desire (your own as much as anyone else's in the audience) to get to wine and cheese and discussion—the real work and fun of the day. That's my position now. I stand, gentle reader, between you and the genial text.

This book falls within a rich new vane in science studies. Van Kerkhoff is a truly engaged scholar—she is arguing for a program of integrated research in which the social scientist works together with scientists, administrators and politicians to make a difference in the world. In her case, a difference which really makes a difference: our ways of managing the world. To the uninitiated, this seems like a no brainer—if we can do good in the world through our studies, of course we should.

Let me tell you a story. As is the way with many theological disputes, the path to this simple statement has been long and tortuous. In the days of the Edinburgh School, many argued that the role of the social scientist was outside of science looking in. In order to accommodate the 'principle of symmetry' we could not choose between different scientific traditions; we were forced methodologically to play the agnostic. There was once some reason for this. It enabled us to say 'it ain't necessarily so', just because the priests of our day (at that time the physicists) declared that they had privileged access to the truth about nature of the universe. We could show that the ways of building truth were the same as ways of building delusion; and that the key to acceptance or rejection of an hypothesis was as much about the range of social, political, organizational work that scientists had to engage in as about its acting as a mirror of nature. At the time, this caused an arguably unfortunate split with the largely socialist-inspired Society for Social Responsibility in Science. The problem with many in that tradition was that they did not see the issue being whether we could know the truth about the world (Marxism in particular failed if it did not deliver unitary truth), but whether the truth could be used wisely. The *nec plus altra* was the cybernetician Wiener arguing—like Comte over a century before—that only the true science of sociology could govern the runaway truths of science. In the act of abnegating its social responsibility, science studies provided a due corrective to the overweening pride of both social and natural scientists—arguing ultimately that both the categories of the social and the natural were fully constructed.

The story so far, then, is that there formed a triune understanding of science: those who believed they had access to natural truth, those who believed they had access to social truth, and those who declared that both emperors were naked. I use this latter image advisedly, since the act of declaring someone naked is quintessentially the act of pointing fingers—not of offering to clothe or recommending a good tailor.

As van Kerkhoff's work amply demonstrates, we are not in a world now where such a split makes sense. Indeed, we really have never been. Recent ecological evidence shows that we as a species have been in the process of killing species and managing the environment ever since our inception. Michel Serres has suggested that it's really only the past two hundred years (since, for example, Malthus—who argued that we had limited agricultural resources pitted against exponential human growth) that we have come to some kind of self-consciousness about the fact. And finally today we have governments hanging in the balance—as demonstrated by the 2007 elections in Australia—according to their policies on human impacts on the environment. We have never been at liberty to ignore the environment—now we are beginning to recognize that fact.

For several years I used the word 'intervention' to describe what I thought we should be doing with scientists. We as social scientists, by this reading, could understand the science better than the scientists, because we understood its organizational, political and cultural imbrication. We could advise them how they should do their own work (along the lines of Latour's *Science in Action* being taken as a primer for training scientists in the Netherlands). However, as Phoebe Sengers pointed out to me—and as van Kerkhoff so ably demonstrates—this is itself an arrogant position. Rather than intervention, we should be thinking engagement, and rather than each retiring to our several citadels we should be thinking integration. It's about time.

And indeed, in this marvelous book, we learn that in many ways it's all about time. Jay Lemke has (mal)adapted the term 'heterochrony' from evolutionary biology to indicate the mismatch between organizational trajectories. My favorite example here is long term ecological research, which when capitalized represents a series of ecological stations funded by the US government to produce very long term baselines of environmental data. The baseline problem is that ecosystems change in, say, 150 year chunks, whereas scientific careers last some 30 years and most ecological data is gathered over the very short term. How to align these? But the problem gets worse—most governments are on, say, five year cycles. Very rarely do they plan over the long term—either socially or environmentally. (The one strong counter example I know here is forestry planning for shipping in Holland in the seventeenth century). One current belief in the United States is that we should let current problems—an aging population, a rapidly changing climate, depletion of petrochemical reserves—wait until they are urgent enough for us to make it economically viable to produce a technofix. Michael Crichton is a populariser of this position with respect to global warming—he argues that we should consume now in order to raise the standard of living and deal with attendant problems down the road when, due to inflation on the one hand and technoscientific progress on the other, it will be cheaper to deal with them. This venal position— *après moi la solution*—is as bad as it's mirror opposite *après moi la deluge*. It is what we are conjuring in the present—and the ways in which we conjure it—that will determine the possibility of solutions, droughts and flood.

In this book, through a wonderfully grounded exploration, van Kerkhoff weaves together stories of integrated environmental management with rich, timely analysis. What more is there to say? Except, *après moi le texte.* Enjoy!

Geoffrey C. Bowker
Cagliari
6 December, 2007

ACKNOWLEDGEMENTS

Of the many people who contributed to the development of this work I would particularly like to extend warm thanks to my supervisors and colleagues, David Dumaresq and Gabriele Bammer. Their variously enthusiastic, critical, and provoking commentaries each contributed not only to the final product, this book, but also to three and a half years of tremendously rewarding intellectual challenge. I am also most grateful to Bill Clark and Geof Bowker who were also greatly supportive and instrumental in getting this work to published form. My thanks also go to my colleagues and friends in the Human Ecology Forum for their constructive critiques, particularly Val Brown and Jacqui Russell for reading earlier drafts of this work in its first incarnation as a doctoral dissertation, and likewise to Roger Clark from the USDA Forest Services for his time and comments.

I must also thank the many case study participants who made this research possible, in particular Roger Shaw and Ian Noble for allowing me into their organizations to ask probing questions. The generosity of everyone to contribute their time made these, in many respects, 'dream' cases. Their willingness to cover part of my expenses to attend their meetings is gratefully acknowledged.

I would also like to acknowledge the support of Land and Water Australia, whose generous scholarship enabled me to focus on the 'real work' of research. More personally, my thanks also go to the staff of Land and Water Australia for their ongoing interest and enthusiasm.

Finally, of course, I must thank my family for their unfailing encouragement. Special thanks to Aleta for her detailed proof-reading and, of course, René for making sure I kept one foot firmly within my social context.

THE CHALLENGE OF
INTEGRATED ENVIRONMENTAL RESEARCH

And making a difference. Yeah, that's it, can we make a difference in the coastal zone? That's a goal. That's all part of bringing people together into something that can make a difference.

Roger Shaw, CEO CRC for Coastal Zone, Estuary and Waterway Management, 2000.

At the turn of the millennium, the role of science in the industrialised and industrialising world is undergoing a curious change. On the one hand, it is widely celebrated, as a source of economic wealth and prosperity, a source of conveniences, security, culture and identity. Yet on the other, it is increasingly questioned—this celebration has been accompanied by a growing awareness of the fallibility of science, for example through the apparent 'failure' of science to solve major environmental issues, such as climate change. Further, as iconic achievements such as Star Wars technology, Dolly the sheep and controversies over stem cell research have become headline news, the professed moral and ethical neutrality of science has been challenged. Both the celebration and the caution can be seen as consequences of a widespread realisation that the boundaries between science and society, carefully nurtured over the last three centuries, are artificial. The industrialised Western scientific ideal of the pursuit of knowledge independent of social constraints and concerns is giving way to a sense that we, collectively, are a fundamentally 'scientific' society, and science is fundamentally social.

While the consequences of these changes are varied, ongoing, and not yet well understood (as I will argue in more detail in Chapter 2), a key theme has been movement away from research that is fragmented intellectually, isolated organisationally and independent of social and economic interests, towards research that is *integrated* across all these dimensions. The realisation of the artificial nature of the boundaries between science and society has paved the way for new relationships that cross the traditional divides. In particular, the belief that scientific credibility is the product of independence and distance from the arenas of policy, economic development, and social change is itself losing credibility. Through engaging in a range of partnerships, researchers are increasingly taking up the challenge of bringing about change, of directly making a difference in the world.

The significance of the shift toward integrated approaches to research for science as a whole is starting to be realised. Taken individually, small concessions away from the scientific ideal of independent, abstract, universal research can be explained without compromising the core distinctiveness that scientists have traditionally relied upon to justify their position in society. Taken collectively, these adaptations represent a major shift—some say a paradigmatic shift—in the philosophy and practice of science (Gibbons et al., 1994; OECD, 1999; Ziman, 2000). Understanding and participating in these shifts while maintaining a sense of what it is to do science constitutes the challenge of integrated research.

While authors have recently begun to document and analyse these changes, to date most analysts have tended to examine the issues surrounding integrated research at the level of the research sector *as a whole*. This study focuses on one area: environmental and natural resource management research (referred to throughout as 'environmental research'—this includes natural resource management research concerned with issues of conservation and ecological sustainability, but not those concerned with a sole focus on production). But first, some further explanation of the general concept of integrated research will be offered.

WHAT IS INTEGRATED RESEARCH?

In general terms integrated research can describe any research that actively crosses conventional boundaries within and around science. It should be noted here that throughout this book a distinction is drawn between science and research. 'Science' refers to the formal, academic, discipline-based institution (including formalised 'social sciences'), whereas 'research' is the open-ended, systematic process of inquiry.

Integrated research commonly appears in two guises. First, there is the integration of *the sciences*. Similar to (and often used interchangeably with) multi-, inter-, or transdisciplinarity, although each of these terms vary in their usage as well, integrated research in this context suggests overcoming the barriers created *within the sciences themselves*. However, other significant barriers to integration of the sciences take the form of organisations (the difficulties of working across different universities and government-funded research agencies, for example), and even within-discipline approaches, such as empirical research and modelling. Calls for integration reflect that science is fractured across many different boundaries.

The second common use of the term integration goes beyond the institution of science to integrate with what might be coarsely labelled as 'non-scientific' institutions. (I use the term 'institutions' to describe large-scale groups of organisations, such as 'science', 'law', 'government', 'industry' and 'community' held together by widely understood formal or informal rules and conventions.) In other words, researchers work actively with policy-makers, community groups, business or industry *in the design, conduct and application of the research*. This represents a significant shift away from the traditional, linear view that science is 'done' by researchers, and then 'applied' by others towards interaction at each stage of the process.

These two uses of the term 'integrated research' suggest activities that can cross any or all of three main sets of boundaries: disciplines, organisations, and institutions. This is not to suggest that such integration is necessarily new—it has been well established in various literatures that science does not take place in a socio-political vacuum, and that any notion of 'pure' science independent of these boundaries is, at best, naïve (Barnes and Edge, 1982; Chalmers, 1982; Latour, 1999; Kates et al., 2001). The newness, rather, is that the connections across these boundaries are being recognised as a legitimate area of scientific participation, rather than as peripheral activity at best, or a design flaw at worst (Gibbons et al., 1994; Irwin, 1995).

Origins of integrated research in the environmental context

As Chapter 2 will elaborate, these changes can in large part be attributed to the dual pressures of the commodification of knowledge (the perception that research is the engine of economic growth), and social rejection of the ideology of impartial, objective science (so that scientific knowledge becomes comparable to any other type of knowledge). The unlikely bedfellows of economic commodification and social critique simultaneously demolished, or at least substantially reduced, the barriers between science and other 'knowledge producers'—that is, everyone else.

Nowotny et al. (2001) describe this new landscape as one in which society is starting to 'speak back' to science. Their analysis suggests that the prevalent model has been for science to only 'speak to' society, and society has listened. In recent years, society is learning to respond. While this characterisation may be appropriate in some parts of the scientific institution, other areas have been struggling to develop a productive conversation with 'society' for far longer than Nowotny and her colleagues suggest. Environmental research is one of those areas.

The historical and contemporary context of environmental research is significantly different from that of the sciences oriented towards enhancing the production of commercial goods. As Bradshaw and Bekoff write: "The unprecedented impact of humans on the land and water worldwide will continue to involve ecologists in conservation, decision and policy-making [sic]. Bosch, Ross et al. (2003) have argued that there is increasing demand from resource managers for integrated solutions to environmental problems. Furthermore, the fact that most environmental problems stimulating current research are products of science and the culture in which they are developed, means that scientists are already deeply involved."(Bradshaw and Bekoff, 2001, p. 460). In other words, environmental research has emerged from social as well as scientific concerns, where science is as much a part of the problem as it is part of the solution. Importantly, this observation applies not only to those countries and cultures that have embraced a scientific-technical mindset, but by virtue of globalisation and global institutions that share and promulgate the ideas of science, such as the World Bank, it also extends now throughout the globe, albeit unevenly and with varying manifestations.

3

While this general level of connection between environmental research and society is important, there are three more specific aspects of environmental management that are significant with respect to integration. The first is the public good nature of most environmental management; the second is its tendency to cross organisational boundaries; and finally, being predominantly issue- or problem-focused, the research that aims to inform environmental management is commonly interdisciplinary.

The 'public good' aspect of many environmental issues is significant in that the relationships between environmental research and society tend to be mediated through government policy, regulation and legislation, rather than through markets. Commercialisation of knowledge for improved environmental management and the creation of artificial markets for environmental services are slowly emerging, but are still highly limited. The majority of environmental management, while it may be the responsibility of landholders or private firms as well as government agencies, is still coordinated largely through policy and regulation. While most analysts have bundled public good and private good research together in their descriptions of integrated research, because both display very similar characteristics and have been subject to similar pressures, the differences between integration motivated by markets and integration motivated by policy interests are yet to be fully explored.

Secondly, because environmental issues tend to be identified according to biophysical characteristics (catchments, coasts, forests, and so on), they also tend to cross managerial or jurisdictional boundaries. Consequently, environmental research has a long history of operating and coordinating across different organisational units—government agencies, land holders, research groups—to achieve improved environmental management.

Finally, the complexity of environmental issues is commonly such that more than one discipline is relevant to the resolution of problems. As many others have noted (Klein, 1990; Board on Sustainable Development Policy Division National Research Council, 1999) working across disciplines has been an ongoing challenge for environmental research, not a new one.

Because of this history, that is shared in significant parts of the health sector and other public research areas, ideas of integration are relatively familiar in environmental research. Further, there is a practical and intellectual history within environmental research of grappling with integration under different guises and in different contexts. This suggests that the implementation and consequences of integration in the area of environmental research are likely to be different from the consequences that may be felt in private good research sectors.

THE CHALLENGES OF INTEGRATED ENVIRONMENTAL RESEARCH

This history of environmental research has left a legacy of many approaches to what is now being formalised as 'integration'. Some of these will be discussed in more detail in Chapter 2; however, in a general sense there is no shortage of methods or tools for doing research that crosses disciplines or involves non-researchers in some way, from participatory methods to action research, from integrated catchment management to adaptive management—and the list goes on. These have been developed and used by researchers over the past three decades to structure their relationships with the non-scientific world.

Yet the point of Nowotny et al.'s work, and that of other commentators on integrated research, is that the non-scientific world itself is changing. The larger forces of integrated research are relevant here: the ever-increasing demand for science is coupled with ever-increasing scepticism and scrutiny by those who would apply it. In other words, the need for science to maintain its credibility is escalating, as is its capacity to be politicised and criticised by both scientists and non-scientists alike. Further, as sources of 'untied' funding diminish and the shortfall is more often being sought in private enterprise or 'interested' sectors of government, claims of scientific impartiality are no longer tenable. As Holling aptly describes the present situation:

> Ecologists are just beginning to develop the range of experience needed to link science, policy and politics. Policy people are largely unfamiliar with ways to recognize the novel interactions now emerging between economic, social and ecological systems. Because the science is in transition, there are not only conflicting voices, there are conflicting modes of inquiry and conflicting criteria for establishing the credibility of a line of argument. (Holling, 1998, p. 1)

The large question of the impact of integrated research on the credibility and reliability of scientific information when the old standards of objectivity, impartiality and disinterestedness can no longer be assumed is being raised in all sectors of research (see for example Huber, 1991; Davidoff et al., 2001; Nowotny et al., 2001). Decision-makers responsible for many forms of environmental management—farmers, policy-makers, urban and rural community groups, business, industry, and many more—are demanding a say in the relationships between themselves and researchers.

Consequently, the activity of integrated research does not begin with the development or application of an integrative model, framework or method. There is a prior step: researchers need to negotiate with those decision-makers *how* the relationship should unfold.

An awkward silence: negotiation, articulation and learning

The importance of negotiation as key to relationships between environmental researchers and non-scientists is not new. In the context of scientists advising environmental policy, for example, Jasanoff (1990) wrote that processes of scientific advisory groups "...are most effective in building consensus and guiding policy when they foster negotiation and compromise" (p. 230). However, negotiating new and innovative relationships between researchers and resource managers around the concept of integration is currently fraught with confusion. The breadth and variation of the term, from overarching 'motherhood' descriptor to highly specific technical models, suggest that the concept of integration is not yet a firm basis for negotiating these relationships (van Kerkhoff, 2005a).

This has led to a situation that can perhaps best be imagined as similar to two sports teams facing each other across a sports field. One team plays soccer, perhaps, and the other plays rugby. Yet they both want to play a game together. How are they going to gain agreement on the rules of the game? This demands two types of action: articulation of the rules of the games with which they are familiar, and negotiation as to how these rules can be adapted to create a mutually satisfying game. This is analogous to the current situation of integrated environmental research. A number of groups are coming together, seeking to work out compatible rules of engagement that will lead to mutually beneficial and satisfying outcomes for all involved. How do they articulate their histories and plans so that they make sense and are relevant to the others? How can ways forward be negotiated?

There has been little research into these issues of articulation and negotiation in the context of integrated research. Yet the quality of *this step* in the development of relationships between science and the decision-making world is crucial for the quality of the eventual actions that are taken. To return to the opening point of this chapter, science has become so heavily entwined in our decision making world *and* our biological world that failure of the system that maintains scientific quality can have dramatic consequences. The risks of taking decisions based on poor science are often noted and discussed, but there are also the risks of taking the wrong action based on poor *relationships* between science and the decision-making (and action-taking) sectors.

There are indications that the 'awkward silence' that can be imagined of the two sports teams in the analogy above is sorely felt in environmental management. As Dovers and Mobbs (1997) note: "Ecology, policy processes and people (communities) have had few connections through which to communicate and work together; they have had no common frameworks for ongoing interaction to move beyond episodic interactions defined by conflict" (p. 39). Bradshaw and Bekoff concur—for example, in their description of the 'science-policy gap' they write: "Such gaps are a type of conceptual no-man's-land, and their significance is underappreciated. Transparency of process is achieved when science gazes upon itself and the interfaces between science, society, humans and the environment are critically examined" (2001, p. 462). This lack of 'transparency' is more than an intellectual curiosity—as Cortner (2000) notes, environmental and natural resource policy is undergoing "significant introspection" (p. 22) as new approaches to the relationships between science and management are tried and tested. The concepts that are available strongly shape what is seen through such processes of introspection, and what can be learnt from it. Those aspects of experience that cannot be articulated remain hidden, obscure forces that are poorly understood and cannot be planned for.

As integrated environmental research is focused on bringing people from different backgrounds together, having concepts that link the groups—and make sense to all of them—is an important component of being able to negotiate how, and how *well*, the connections between them are made. The absence of clear concepts has not stopped integrated research from going ahead—people are grappling with these first steps, and establishing good (and not so good) relationships. However, the lack of transparency, lack of ways of abstracting from the immediacy of day-to-day events, mean that these experiences are seldom a source of learning for those not immediately involved.

As later chapters will illustrate, without a broader theoretical, or abstract conceptual structure to give their experiences meaning, people involved in integrated research commonly regard their experiences as irrelevant, too context-specific, too *personal*, to offer other researchers any useful insights. Alternatively, where they do believe they have insights that would be useful to others, the problem of abstraction remains. How much of the events they have participated in were due to the specific context (individuals involved, exigencies of place, vagaries of political climate and public support...), and how much can be attributed to a broader, shared context of integrated environmental research that can offer useful lessons to other people in other situations? The desire for transparency, then, can also be understood as a desire to learn from others, to avoid having to reinvent the wheel every time a new integrated research endeavour is embarked upon. Bammer (2005) has argued that this needs to become recognised and developed as a new specialised skill. To achieve this, the 'conceptual no-man's-land' needs to be claimed and populated with ideas and concepts. By articulating some concepts, based on the systematic study of 'those who have been there'—even if those concepts themselves become hotly contested—the prospects for future relationships within integrated environmental research are surely brighter.

THIS STUDY

The science studies context

This study is located within the field of science studies. Science studies takes many forms, and has become a popular proxy term for a wide range of sociological and anthropological approaches to research on science. However, this study differs from the majority of science studies in two main ways.

First, it is not *primarily* concerned with those aspects of scientific practice in which scientists construct facts or engage with the objects of their research. This study, being concerned with integration within and beyond science, focuses on those aspects of scientific practice that are concerned with relationships among people, and how they articulate their problems and achievements of integrated research.

Second, this study differs from the majority of counterparts in science studies in that it is based on a methodology that seeks to *facilitate the development* of research practice. This is a significant departure from science studies (and other social research areas, such as sociology of science, or sociology of scientific knowledge) that is primarily concerned with the evaluation and critique of scientific practice, or with building theory of science that is relevant to science studies but does not seek to engage scientific practice. As such, while the investigation and theoretical development of integrated environmental research is the main theme of this study, the methodology used to do that forms a second strand of theoretical development.

The second strand is concerned with the development of an approach to science studies premised on engaging scientists in mutually beneficial research, as a practical way of moving beyond what has now become widely known as 'the Science Wars'. The Science Wars will be discussed in more detail in Chapter 2, but in brief the term encapsulates an ongoing series of exchanges between social studies of science (including sociology and science studies) and biophysical scientists. These have, in many instances, become quite acrimonious, and empha-sised to the scientific community at large the critical aspects of social studies of science. Yet there is clearly potential for science studies to engage *with* scientists more productively, and this can be to the mutual benefit of science studies and science more broadly. As Jasanoff has noted: "To rebuild the connections between S&TS [science and technology studies] and its potential audiences and allies, we will first have to rediscover how to talk to practitioners of science, engineering and medicine in ways that are seen as enlightening and fruitful rather than as dismissive or demeaning" (Jasanoff, 1998, p. 94). However, despite several calls for the relationship between the warring parties to move on into more harmonious and collaborative modes of operation, as yet science studies has few methodological approaches that are premised on mutual benefit or shared learning.

One of the key issues here is that of the position of science studies researchers, and their relationships with those they are working with. Science studies research-ers who study the relationships of *others*, with a goal of mutual benefit, must also be significantly introspective about the ways they construe *their own* relationships with their study participants. This issue is central to the way this study was designed and conducted, and will be briefly described here.

This research was based on a research methodology that draws on an established science studies approach, that can be loosely clustered around the title 'research as practice' (Lave and Wenger, 1991; Pickering, 1992; Chaiklin and Lave, 1993; Pickering, 1995; Lave, 1996; Wenger, 1998). Broadly speaking this is an ethno-graphic approach that focuses attention on the day-to-day activities of the research-ers engaged in integrated research. However, 'practice' is an extraordinarily diverse concept. To narrow the diversity, and to provide a methodological framework within which the researcher's relationships with the study participants becomes central, the notion of 'practice' is interwoven with a similar, albeit largely separate approach based on social approaches to communication (Sless, 1986; Leeds-Hurwitz, 1995; Stewart, 1996; Penman, 2000). While there are many variations, in general social approaches to communication are based on a sense of communication as a shared event or activity in which meaning is mutually co-constructed. In the context of research, social communication approaches direct attention to the relationship between researcher and participant as an ongoing collaborative engagement, emphasising shared appreciation and learning. The emphasis on social communication within a broader framework of practice led me to title this approach 'social communicative practice'.

Social communicative practice offers one step on a path towards 'fruitful' engagement between science studies and scientists. It focuses our attention on the ways in which people articulate and negotiate meaning, and come to understandings that form a basis for action.

Overview of the study

Given the broad *problematique* of integrated research and the science studies context, this study had four aims.

The first was to develop an account of the practices of integrated environmental research 'from the trenches', so to speak. How do people—not only researchers, but all participants in integrated research—make sense of and articulate their experiences of working together? How are joint decisions made? How are problems negotiated? Creating a current account of integrated environmental research that draws upon *the way people experience it* provides a basis for theoretical development that is relevant to practice.

The second aim of the study was to contribute to the development of conceptual tools that can be used to better articulate the activities of doing integrated research, and thereby contribute to how others can learn from them. The purpose of these tools is to help differentiate the systemic features (in the sense of those features that are shared and related to each other) of integrated environmental research from the idiosyncratic. How can new participants learn from the processes they or others have engaged in?

Associated with that was a third aim: to contribute to the philosophical understanding of integrated environmental research (and, perhaps, integrated research more generally), and how it differs from conventional research. This was less concerned with practical tools, but rather with taking steps towards defining ways of thinking about science that might help to relate the practice of integrated research to conventional views of the practice of science. How is integrated research different from conventional research? How is it the same? How do these characteristics relate to each other? Given the conceptual no-man's-land noted in the previous section, it would be hoped that this would also contribute to the second aim.

Finally, as discussed in the previous section, the fourth aim of this study was to develop and apply a methodology that combined research as practice and social communication approaches as a basis for theory development that is relevant to both science studies and science.

This research consisted of two case studies, each Australian environment sector Cooperative Research Centres (CRCs). CRCs are substantial research organisations, most with around 50 full-time equivalent staff, and annual budgets in the order of AU$8-12 million. They are funded jointly through the Australian Federal Government and the research partners that make up the Centres: universities, government researchers, industry, and other government agencies. The first CRCs began operations in 1991, and in 2001 there were 64 Centres in operation in Australia (CRC Program, 2002). As will be discussed in Chapter 3, their funding rules require that CRCs engage in—and demonstrate—'integrated research'. Consequently they offer a wealth of experience of the joys and traumas of integrated research practice that is certainly unique within Australia, and has few parallels internationally.

The requirement that integration be demonstrated, without many strictures on how to do that, has also created an environment where new CRCs have considerable leeway to interpret the notion of integrated research as they see fit. Consequently this study was able to compare how two different Centres, working within the same general research structure, 'created' understandings of integrated research that suited vastly different socio-political contexts. One Centre, the CRC for Greenhouse Accounting, was operating within a volatile international and national political context created by the Kyoto Protocol negotiations. The other, the CRC for Coastal Zone, Estuary and Waterway Management, was working in a highly dispersed political situation, with many different authorities responsible for coastal management, as well as considerable community and industry interest. The potential for these different contexts to influence the ways in which the Centres' research was integrated created an ideal milieu within which to explore the negotiation and articulation of the concept of integrated research.

The study was carried out with both organisations over a period of approximately 18 months from mid-2000 to early 2002. While the methods used (primarily in-depth interviews, participant observation and document analysis) were fairly typical of qualitative research, in this study they were placed within a wider methodological framework that characterised engagement in research as participation in diverse, iterative, ongoing conversations. These conversations, as interaction between myself and the study participants, formed the basic unit of analysis within the structure of research as social communicative practice. I conducted 40 interviews (plus 6 as a pilot study) over two rounds roughly 12 months apart. In the second round of interviews some findings from the first round were presented back to participants as a basis for discussion, and key issues that emerged from the first round were explored in more detail. The interviews were transcribed and analysed using N*Vivo qualitative data analysis software. I also attended two annual conferences for each of the CRCs, which ran for four days each, as a participant observer. Field notes from these meetings were integrated into the analysis. I also analysed a wide range of documents. The analytic techniques I used included line-by-line micro-analysis and open coding (Ezzy, 2002; Strauss and Corbin, 1998).

There are, of course, many areas this study will not cover in detail. In particular, as the notion of integrated environmental research is concerned with the practices of building and maintaining relationships among people, many other aspects of organisational or scientific life appear only incidentally to these concerns. Importantly, this study was not directly concerned with *conventional* aspects of scientific research, such as the relationships of researchers to the biophysical world, scientific instruments or computing, or those individual aspects of scientific work, such as inspiration or curiosity. It should be stressed here that this is not because these aspects are perceived as unimportant or impossible figments of a socially constructed imagination, a charge that is commonly levelled at those who choose to study human–human relationships within science in preference to human–biosphere or human–machine (see, for example, Pickering, 1995, and Latour, 1999, especially Chapter 1). Rather, these aspects are regarded here as elements of the traditional practices of research, and were bracketed aside to allow the 'integrative' aspects to take centre stage.

Other issues arose according to the importance they were attributed in the conversations that are the basis of this study, and according to their relevance to the aims of the study. These include (but are not limited to) issues of organisational structure, power relations, organisational or institutional culture, and scientific practice.

Each of these factors are weighty topics in their own right, and do emerge at different points. However, the point of the focus on conversations is that such topics emerge *as the participants saw them*, in dialogue with my own aims regarding integrated research.

Organisation of the book

The eight chapters of this book correspond roughly to two main sections. This chapter and Chapter 2 give the background to the study. Chapter 2 gives an overview of three different literatures that illustrate the unique history of the concept of integration in environmental research, and highlight the tensions generated by different, potentially conflicting forces of integration. Chapter 3 details the organisational and political background of the case studies.

Chapters 4 through 8 build a theoretical conception of integrated environmental research, based on the case study material. The structure of the second half of the study builds a conceptual picture of integrated environmental research by starting with themes that emerged from the study, and using those as a base to progressively build up more synthetic, theoretical ideas. It is through this process of theorising from practice that new understandings of integrated environmental research are proposed. Chapters 4 and 5 focus on the extent to which participants' *understandings* of integrated research (reported in Chapter 4) captured their *experiences* of the activity of integrated research (given in Chapter 5). In Chapter 6 I explore how people were able to create and maintain relationships *despite* the gaps, tensions or conflicts between understandings and practice. In Chapter 7 I draw out some of the key sense-making constructs reported in Chapter 6, and develop some theoretical concepts that can be used to abstract from the immediacy of experience, so that they may form a basis for articulation and negotiation of integrated research. Chapter 8 develops the theoretical implications of these differing views for the relationship between integrated research and conventional science.

These understandings are offered as a contribution towards the articulation and negotiation of integrated research, and so may help generate research that is better equipped to make a difference.

THE FRACTURED LANDSCAPE OF INTEGRATED RESEARCH

...it must be recognized that the relationship between scientific research, education, technological innovation and practical benefits is much more diverse and complex today than in the past, and frequently involves many players other than researchers. The progress of science cannot be justified purely in terms of search for knowledge. In addition, it must be defended... through its relevance and effectiveness in addressing the needs and expectations of our societies. (UNESCO, 1999)

In Chapter 1 it was argued that the idea of integrated research has emerged to counter the fractures and disjunctures among disciplines and organisations of science, and between science and other institutions. These fractures are widely held to inhibit the accessibility and usefulness of research to decision-makers and action-takers. Yet the pressures for integration are not quite as straightforward as the aim of increasing the relevance of research to users might suggest. The pressures for integration are themselves fractured into at least two ideological camps, with a third ideology resisting integration. These ideologies represent different understandings of knowledge, how it takes its 'value' in society, and its relationship to action. Consequently, it is not only that integration takes place against a fragmented background of disciplines, organisations and institution that renders the idea of integration complex and multi-faceted—it is also that the *arguments for* integration are in conflict, and are controverted by *arguments against* integration.

One argument for integration, the knowledge economy, is a description in common usage. I have labelled the other argument for integration the 'knowledge democracy', and the argument against integration as the 'knowledge autocracy', in an adaptation of Jasanoff's (1990) 'democratic' and 'technocratic' paradigms of public policy decision-making. Of course, these are not the only arguments for or against integration, but do underlie much of the complexity and confusion surrounding the idea of integration.

I have used these three categories to structure my description of the fractured landscape of integrated research in this chapter. Within each section, the basis of the argument for or against integration will be summarised. The summary will be followed by an overview of academic literatures that draw on that argument, in theoretical and practical models of integration. Each section will close with a brief sample of policies that exemplify the ways the different arguments are used in practice.

While this allows me to draw connections between several largely disparate areas of literature in a coarse-grained way, the cost of this strategy is, of course, detail. Some parts of these literatures that are particularly relevant or indicative of the larger themes will be featured and discussed in more detail, with a consequent sense of 'zooming in' and 'zooming out'. While those aspects that are zoomed in on are relevant in their own right, I acknowledge that they are not the only possible candidates for detailed examination. Consequently readers will inevitably see different aspects of their own areas of expertise or interest that could (or should, perhaps) be included or covered more fully.

THE KNOWLEDGE DEMOCRACY

The knowledge democracy is an argument for integration between researchers and other 'publics'. It is based on the view that including lay citizens and dispersed interest groups is both ethical and practical as it will help to ensure that research is targeted towards the needs of the communities it affects. It is *democratic* in the sense that it seeks to assert that the knowledge of participants from outside the scientific institution is valuable and can be combined usefully with scientific knowledge. This argument is prominent in environmental research, where it has emerged largely from reactions against agricultural and development situations where those involved in managing the natural resources were excluded from research about those resources, leading to inappropriate research that was often of little use or, occasionally, had serious negative consequences.

Environmental context and theory: from extension to participation

The theoretical and practical developments in integrated environmental research are closely allied with developments in agricultural extension; indeed, to the extent that agricultural management is concerned with the care and stewardship of natural resources, they merge (Buttel, 1992). Consequently, many of the ideas behind agricultural extension can be seen to underpin much of the current thinking on integrated environmental research.

Agriculture and science   The emergence of modern-day industrial agriculture has been closely allied with science. Rossiter (1975), for example, describes in detail the development and use of agricultural chemistry and soils science from the 1830s in the United States, stating that "…few doubted in the 1830s and 1840s that chemistry could solve the problems of agriculture" (p. 11). The relationship between scientists and landholders was built further in the US, with the establishment of the Land Grant Universities in 1862, under the clear mandate to broaden access to technical knowledge (Brannon, 2002).

The knowledge democracy emerged in reaction against the traditional 'extension' or 'transfer of technology' model that underpinned the relationships between researchers and land managers. The transfer of technology model is essentially a linear research and application process, similar to the one depicted in Figure 1.

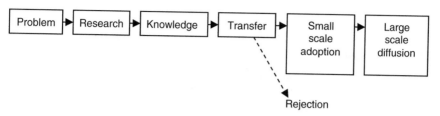

Figure 1. The linear model of extension.

Scientists set the research agenda, do the research, and then 'transfer' the results to the farmers. The results then 'diffuse' through the farming community. As Scoones and Thompson (1994) write:

> The superiority of 'rational science' is assumed and the pursuit of change (development) is derived almost exclusively from the findings of the research station and transmitted to the farmer through hierarchical, technically oriented extension services. Farmers are seen as either 'adopters' or 'rejectors' of technologies, but not as originators of either technical knowledge or improved practice. (p. 18)

As such, the traditional extension model is positivistic, in that there is assumed to be an objective truth that the scientists pass on to the farmers via extension officers; and individualistic, as farmers are assumed to make decisions independently on a rational basis. It is perhaps unsurprising then that although this form of extension was effective in the dissemination of new technologies for increased production, it was far less successful in convincing farmers to prevent or ameliorate land degradation (Ison and Russell, 2000).

Also important in the rise of the knowledge democracy were the perceptions that vested interests were promoting research and its associated products (fertilizers, pesticides, and so forth) that were not in the best interests either of the farmer or for the maintenance of their land. This was particularly the case as so-called 'Western' technologies were heavily promoted in developing countries, where farmers undertook massive debts to use the technologies being sold by companies based in industrialised countries, and applied them in situations that were vastly different from those they had been designed for (see Glaeser, 1987, for an overview). The failure of the Green Revolution in many developing areas and the social and economic impacts it had on poor communities—especially those who were already marginalised, such as women or indigenous groups—led to a strong critique of the transfer of technology model, from the point of view of human rights and welfare as well as its impacts on production. However, these critiques were not restricted to developing countries. In industrialised countries the dominance of production-based agricultural technologies at the expense of land conservation was also questioned for its ecological, economic and social impacts (for Australian examples, see Barr and Cary, 1992; Vanclay and Lawrence, 1995). Thus a widespread rebellion against the transfer of technology extension model emerged.

In the 1990s, participatory research became a popular alternative to 'top-down' extension. Scoones and Thompson (1994) claim that participatory research approaches in general:

> emphasize the rational nature and sophistication of rural people's knowledge and believe that their knowledge can be blended with or incorporated into formal scientific knowledge systems. [I]f local knowledge and capacities are granted legitimacy within scientific and development communities, existing research and extension services will pay greater attention to the priorities, needs and capacities of rural people and, in the end, achieve more effective and lasting results (p. 18).

As such, participatory research asserted that different types of knowledge could and should be compatible and complementary and actively combined for improved natural resource management. Some of the major research approaches that use participatory methods are listed in Table 1, although there are many more.

Table 1. Research approaches based on participatory methods

Approach	Major references
Farmer-first	Chambers et al., 1989; Cernea and World Bank, 1991; Chambers, 1997
Rapid Rural Appraisal/ Participatory Rural Appraisal	Chambers, 1980; Ampt and Ison, 1989
Agroecosystems analysis	Conway, 1987
Farming systems research	Tripp, 1992
Participatory action research	Fals-Borda and Rahman, 1991; Rahman, 1993
Farmer participatory research	Okali et al., 1994
Participatory technology development	Chambers and Jiggins, 1986
Second order research	Ison and Russell, 2000

Participatory approaches have become popular within agricultural extension, particularly in the context of engaging land managers in efforts to improve the sustainability of their operations. Keen (1997) has usefully summarised the main catalysts for shifts towards participatory approaches in what she broadly terms 'land management research', as illustrated in Figure 2.

This diversity of pressures for change, from academic and community, as well as policy and research funding bodies, illustrate the various strengths of the participatory approach. In theory, participatory research solved many of the problems of the transfer of technology model. It reduced perceptions of neo-colonialism in developing countries, it gave farmers and landholders input into the definitions of problems, and therefore made solutions more viable in the context within which they would be applied. Finally, it reduced the risks that community groups or non-government organisations would reject decisions based on research they had been a part of (For a critique of participatory approaches, see Cooke and Kothari, 2001).

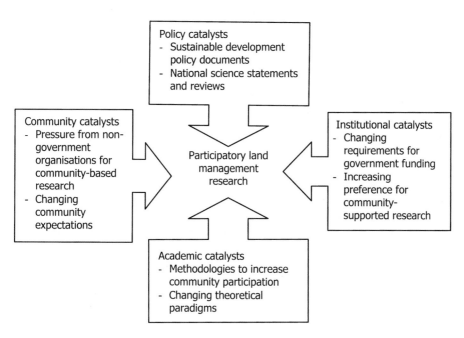

Figure 2. Catalysts for change towards participatory research.
(Source: Keen, 1997, p. 90)

Participatory approaches were important in the development of the concept of integrated research, as they asserted that integration of different knowledges and interests between science and non-science sectors was both epistemologically possible and morally desirable (Scoones and Thompson, 1994; Ison and Russell, 2000). They also highlighted the political nature of research in general, and argued that the distribution of power in conventional transfer of technology models was distinctly masculine and hegemonic, and this was a source of its moral and practical failings (see, for example, Chambers, 1980; Chambers, 1983; Pretty et al., 1993; Freire, 1996).

Hence participatory research can be regarded at a general level as the vanguard of the knowledge democracy in research practice.

Models of 'knowledge democracy' integration

While agricultural extension models of participation tended to focus on involving community groups and landholders in research, other 'democratic' models also emerged that emphasised integrating research with government, or both.

Involving government: adaptive management   Other models of integration were concerned with integration between researchers and policy-makers, rather than community. One of these was Adaptive Environmental Assessment and Monitoring (Holling, 1978), an influential model that spawned several variations that can be more loosely grouped under the title 'adaptive management'. Adaptive management models and approaches drew on systems theory to suggest that policy interventions, or environmental management more broadly, should be regarded as experiments, with concomitant assessment and monitoring as a basis for ongoing learning (Holling, 1978; Lee, 1993; Gunderson et al., 1995).

This was a major step in the environmental research literature towards integration. Dovers and Mobbs (1997), for example, describe the important features of adaptive management approaches as "information is central, the focus is on integrating natural system and institutional/social dimensions, and it is absolutely and inevitably interdisciplinary" (p. 43). In comparison with participatory approaches, which tended to focus on integration at the grassroots-community level, adaptive management focused on integration between science and policy, thus bringing government-level decision-making to the fore. This work was significant because it insisted that researchers and managers could work together in productive, ongoing relationships, in which the science and the management activities dovetailed and strengthened each other. Later versions of adaptive management (for example, Lee, 1993), have productively combined the more quantitative dimensions of adaptive management with social and organisational learning concepts, broadening the integrative scope of the models to include social factors.

In their insistence on partnerships, adaptive management models have contributed to the democratisation of knowledge in their acknowledgement that government agencies also needed to be included as legitimate 'knowers'. In addition, by including government mandates, adaptive management could also link science with broader democratic processes.

Involving both: Integrated... Management   Over a similar period of time as the emergence of participatory research and adaptive management, there also emerged three 'integrated management' approaches. While participatory approaches focused on community level relationships between research and land managers, and adaptive management was particularly appropriate for relationships between researchers and government managers, other 'integrative management' approaches aspired to do both.

Integrated Catchment Management (ICM) emerged in Australia in the late 1970s. By the mid-1990s the concept of ICM (or close variations of it) had been implemented in policy or legislation by every Australian State government (Bellamy et al., 2002). ICM attempts to bring together all the relevant players with an interest in managing a catchment (or watershed), including scientists, community groups, landholders and government agencies at all levels, to develop a more holistic management approach (Hinchcliffe et al., 1999).

This was a significant step in the development of the concept of integration in environmental research and management for several reasons. First, it strongly encouraged the integration of scientific research to address a whole catchment. As Bellamy et al. (2002) note: "Researching and integrating scientific knowledge has been an underpinning principle of ICM processes in Australia." (p. viii). However, it also integrated the science with landholders—'integrated' meant explicitly including the people who lived in the catchment, and their actions. In this way, ICM was probably the most recognised fore-runner of integrated research as it is applied in environmental research in Australia in the late 1990s.

Integrated Resource Management (IRM) was proposed as a broader alternative to ICM in the mid-1980s (Mitchell, 1987). IRM once again broadened and slightly shifted the idea of integration, recognising that while ICM focused on water, similar, more abstract principles could be applied to other natural resources, such as forests or coasts. There has been considerable enthusiasm—indeed, high expectations—within Australia about what can be achieved through IRM, as Bellamy et al. (1999) write: "The basic assumption that IRM approaches contribute effectively to the better management of our land and water resources has not been systematically or critically addressed" (p. 338). A decade later Margerum and Born (1995) proposed the concept of 'Integrated Environmental Management' as more inclusive again, aiming to integrate all human activities in a defined environmental system.

As these examples illustrate, the formal use of the term 'integration' has an important history in land management and environmental research. It has been used in the sense of researchers needing to actively work outside their conventional disciplines and in terms of forging relationships with non-scientists. Moreover, integration has strong associations with the knowledge democracy through the emphasis of Integrated Resource Management and Integrated Catchment Management on community participation and government involvement in research. This embeds the concept of integration firmly within an epistemological and moral framework that rejects the hegemony of science and scientific knowledge, and supports the inclusion of different ways of knowing into environmental decision-making and land management.

However, these issues are not restricted to environmental and agricultural sectors. Authors such as Irwin argue that these areas are only the more visible symptoms of a condition that affects science as a whole. In his book *Citizen science*, Irwin (1995) argues that there is a need for science more broadly to take the needs, concerns and perspectives of society into account, as:

the alternatives are either to argue that the current relationship between science and citizens is unproblematic and therefore does not require modification (a conclusion which is disputed by all the evidence in this book) or to deny that science should have everyday relevance (which will inevitably lead to an even greater public onslaught on science). (p. 167)

This view has been echoed in concerns expressed in the highest echelons of science (Lubchenco, 1998).

Post-normal science Other democratic models of integration were concerned with science writ large. Post-normal science was essentially an early recognition of the need for new types of research in response to different problem–solution situations. Driven in large part by environmental issues, but not restricted to them, the authors of post-normal science, Funtowicz and Ravetz (1993) described the differences in approach needed when the decisions requiring scientific input were both highly uncertain, and had major consequences if the scientific and policy 'answers' or solutions were wrong. As Ravetz (1999) described it:

Going beyond the traditional assumptions that science is both certain and value-free, [post-normal science] makes 'systems uncertainties' and 'decision stakes' the essential elements of its analysis. … Its theoretical core is the task of quality assurance; it argues the need for new methods, involving 'extended peer communities' who deploy 'extended facts' and take an active part in the solution of their problems. (p. 647)

In other words, situations where "typically facts are uncertain, values in dispute, stakes high and decisions urgent" are classified and placed in a systemic relationship to traditional scientific endeavour, rather than being assumed to be rare, aberrant or just not science. This is illustrated in Figure 3 below.

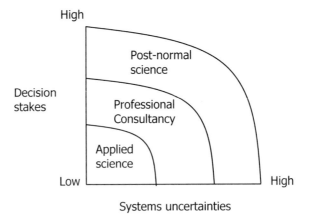

Figure 3. Post-normal science. (Source: Ravetz, 1999)

In terms of process, post-normal science calls for the broader participation of stakeholders, although the authors caution that this is not necessarily for the pursuit of democratic values. Rather it is to expose the decision-making processes to as many forms of quality assurance as necessary to reach a 'good' decision, not just those of scientific methods and peer review. This involves structures to incorporate both non-scientific values and non-scientific, 'extended' facts.

As such, integration is a process or strategy for dealing with the situation the model describes. It is an effort to create a conceptual space that spans the boundaries between science and 'society', where there is "...a recognition that no side necessarily has a monopoly of truth or morality." (Ravetz, 1999, p. 653). However, while their focus on risk and values is extremely broad in one sense, it is also limiting, in that many other decision-making scenarios also require (or may benefit from) better incorporation of scientific and non-scientific information. Consequently, post-normal science offers some much-needed insight, structure, and legitimisation for doing research based on integrating different knowledges.

Sustainability science At a different scale, the concept of 'sustainability science' was proposed as an alternative framework to conventional research. Concerned that science has become 'estranged' from the sustainable development agenda, the authors who coined the phrase (Board on Sustainable Development, Policy Division National Research Council, 1999; Kates et al., 2001) argue that science needs to be done differently if it is to reclaim a key role in environmental decision-making.

Sustainability science demonstrates the ambiguity of the role of the scientist: being at once an analyst *of* nature–society interactions, and a key player *within* those interactions. While acknowledging—even emphasising—the importance of incorporating other types of knowledge and describing participatory approaches as "critically needed" (Kates et al., 2001, p. 641), most of the seven 'core questions of sustainability science' suggest a fairly conventional scientific perspective, with the exception of the last. These questions are summarised in Table 2. While each of these questions is *about* nature-society interactions, only the last question suggests that the researchers may be an integral part of this system, rather than being distant, external examiners of it. Each contributes to the characterisation of the scientific expert as being the one who will ultimately provide 'the answers'.

But more significant than *what* they propose sustainability science is about, is *how* they propose it should be done. As the last question of Table 2 indicates, sustainability science is based on an adaptive management approach. It promotes "societal learning" and a form of action research, where "scientific exploration and application must occur simultaneously" (Kates et al., 2001, p. 641).

In other words, these questions represent a systemic approach, with an underlying attempt to grapple with the need to integrate other knowledge and to work with policy-makers to understand the application of science and its impacts. But this integration appears to fall short of granting *all* non-scientists legitimacy in terms of knowledge and knowing. The following passage is quoted at length, as it outlines how the concept of integration is invoked:

Table 2. Core questions of sustainability science. (Source: Kates et al., 2001, p. 641)

Core questions of sustainability science
How can the dynamic interactions between nature and society... be better incorporated into emerging models and conceptualizations that integrate the Earth system, human development, and sustainability?
How are long-term trends in environment and development... reshaping nature–society interactions in ways relevant to sustainability?
What determines the vulnerability or resilience of the nature–society system in particular kinds of places and for particular types of ecosystems and human livelihoods?
Can scientifically meaningful "limits" or "boundaries" be defined... beyond which the nature–society systems incur a significantly increased risk of serious degradation?
What systems of incentive structures—including markets, rules, norms, and scientific information—can... improve social capacity to guide interactions between nature and society toward more sustainable trajectories?
How can today's operational systems for monitoring and reporting on environmental and social conditions be integrated or extended to provide more useful guidance for efforts to navigate a transition toward sustainability?
How can today's relatively independent activities of research planning, monitoring, assessment, and decision support be better integrated into systems for adaptive management and societal learning?

Sustainability science will therefore have to be above all else integrative science—science committed to bridging barriers that separate traditional modes of inquiry. In particular, it will need to integrate across the discipline-based branches of relevant research described above—geophysical, biological, social, and technological. The same can be said for sectoral approaches that continue to treat such interconnected human activities as energy, agriculture, habitation, and transportation separately. In addition, sustainability science will need to integrate across geographic scales to eliminate the sometimes convenient but ultimately artificial distinctions between global and local perspectives. Finally, it will need to integrate across styles of knowledge creation, bridging the gulf that separates the detached practice of scholarship from the engaged practice of engineering and management. (Board on Sustainable Development, Policy Division National Research Council, 2000, p. 283)

While engineers and the potentially very broad category of 'managers' may therefore be included in the integration of sustainability science, there is some disjuncture between that and the need to involve: "scientists, stakeholders, advocates, active citizens, and users of knowledge..." (Kates et al., 2001, p. 641).

In summary then, to varying degrees, participatory approaches represent an ontological shift away from the idea of a single-reality world, and an epistemological shift away from the notion of science as the sole means of uncovering 'truth' about the world. More radical versions of participatory research claim a democratic, emancipatory ethos, and promote participation as an alternative ethical and political framework to that of conventional science. However, as the last example of this section, 'Sustainability Science' showed, these are by no means simple or unambiguous shifts for researchers to embrace. There are conflicts and uncertainties that are generated, especially with respect to conventional views of science.

Examples of the knowledge democracy in environmental policy

The mere existence of research models based on the knowledge democracy does not, of course, ensure their uptake and use. However, the ideology of the knowledge democracy and its connection with 'integration' is also evident in environmental and agricultural policy at all levels. Some major international and national examples are given here.

Agenda 21   Agenda 21 (UNCED, 1992) was one of the key documents to emerge from the United Nations Conference on Environment and Development held in 1992. Hailed at the time as a watershed for environment and development planning, Agenda 21 has since formed a policy basis in many environmental management contexts worldwide. Integration was a core theme of Agenda 21, prominent in its major argument for the greater integration of indigenous and scientific knowledge in planning and decision-making. For example:

> One of the fundamental prerequisites for the achievement of sustainable development is broad public participation in decision-making. Furthermore, in the more specific context of environment and development, the need for new forms of participation has emerged. This includes the need of individuals, groups and organizations to participate in environmental impact assessment procedures and to know about and participate in decisions, particularly those which potentially affect the communities in which they live and work. (UNCED, 1992, Section 23.2)

Chapter 35 of Agenda 21, entitled 'Science for sustainable development', further strengthens this perspective with respect to research. Table 3 selectively summarises the text of the program area "Strengthening the scientific basis for sustainable management" to illustrate the emphasis on the knowledge democracy, especially the inclusion of traditional and indigenous knowledge.

Agenda 21 therefore strongly supported the 'knowledge democracy' view of integration, and emphasised the role science needed to play in integrating not only different types of knowledge from within the sciences, but also in bringing traditional knowledge into broader decision-making frameworks.

Table 3. 'Integration' in Agenda 21. (Source: Chapter 35: 'Strengthening the scientific basis for sustainable management', UNCED, 1992)

Objectives	– widening of the scientific base and strengthening of scientific and research capacities and capabilities; – policy formulation, building upon the best scientific knowledge and assessments; – The interaction between the sciences and decision-making; – The generation and application of knowledge, especially indigenous and local knowledge, to the capacities of different environments and cultures – Improving cooperation between scientists by promoting inter-disciplinary research; – Participation of people in setting priorities and in decision-making relating to sustainable development.
Activities	Develop methods to link the findings of the established sciences with the indigenous knowledge of different cultures. ... They should be developed at the local level and should concentrate on the links between the traditional knowledge of indigenous groups and corresponding, current "advanced science", with particular focus on disseminating and applying the results to environmental protection and sustainable development.
Implementation	Supporting new scientific research programmes, including their socio-economic and human aspects, at the community, national, subregional, regional and global levels, to complement and encourage synergies between traditional and conventional scientific knowledge and practices and strengthening interdisciplinary research related to environmental degradation and rehabilitation.

These principles have been incorporated into national policies in Australia, such as the National Strategy for Ecologically Sustainable Development, which states: "decisions and actions should provide for broad community involvement on issues which affect them" (Commonwealth of Australia, 1992). They have also been widely taken up and adapted throughout Australia via Local Agenda 21 programs. These are managed and implemented at the municipal government scale, in recognition that local authorities are often in the best position to bring about direct changes in policy, planning, and actions.

Land and Water Australia
The arguments of the knowledge democracy are also evident in other domestic Australian environmental or natural resource management research policy. Land and Water Australia is a research and development funding body financed by the Australian Commonwealth Government. In their Strategic Plan of 2000–2006, Land and Water Australia supplemented their largely biophysical research funding structure around two sets of 'integrative themes' and appointed an 'Integration Manager' to oversee how the pieces came together. The Plan states that:

A perennial challenge for most agencies in natural resource management is that of integration: across issues and programs: across different scales of activity: across different jurisdictions: across ecological, economic and social factors; and across the spectrum from knowledge generation to its trans-formation and utilisation. [...] It is much easier to deliver on neatly defined, discrete (and discreet) research agendas within a single discipline with a common language and worldview than to attempt integrated approaches. (Land and Water Australia, 2001, p. 19)

Their approach to this challenge is via a matrix of themes and issues, represented visually as in Figure 4. In this construct the biophysical themes of 'Sustainable Primary Industries' and so forth are surrounded by human activities such as 'learning and understanding'. These 'human' themes are described as "...powerful lenses through which to examine, analyse and interpret R&D outputs across our entire portfolio" (p. 21). This plan recognised that natural resource management is as much to do with activities of understanding, valuing, learning and so forth as they are to do with the biophysical resources themselves, and that a key part of the challenge of integration is to bring those dimensions into their research planning and funding.

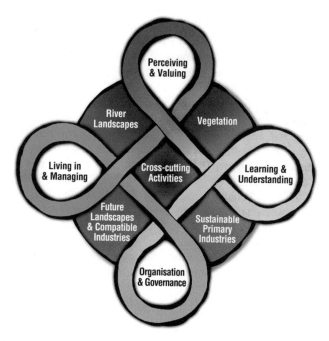

Figure 4. Integrating themes and issues. (Source: Land and Water Australia, 2001, p. 20)

While Land and Water Australia are a relatively small funding body, they are well known for their innovative approaches to research and have considerable influence on the environmental/natural resource management research communities in Australia. Their emphasis on integration and the knowledge democracy is a significant symbol in Australian environmental research, and helped to foster innovative research topics and approaches.

The knowledge democracy and social studies of science

Although the knowledge democracy has been prominent in environmental research, there have also been important 'democratic' arguments from the academic disciplines and sub-disciplines engaged in what might be broadly called 'social studies of science', including history of science, sociology, and anthropological studies. Generally speaking, social studies of science are less concerned with the practical outcomes of democratising knowledge than their environmental counterparts, but do make important theoretical contributions to the argument of the knowledge democracy.

Kuhn and the history of science   The philosophical debate surrounding the nature of scientific 'truth' and its relationship with society and other kinds of knowledge is hardly new territory. Yet while these longer histories are acknowledged, I will (as many others have done) take the seminal writings of Thomas S. Kuhn as a starting point. The publication of Kuhn's historical work *The structure of scientific revolutions* in 1962 (1962/1970) represented a turning point in the formal study of history and sociology of science, what Sardar (2000) has called "…a new phase in the ideology of science" (p. 34).

However, Kuhn's writing was not a revolution in isolation—the sociology of science was emerging as a recognised field at the same time, with the publication of *The sociology of science* in 1962 (Barber and Hirsch, 1962). While Karl Popper's theory of falsification was a prominent topic of debate in the philosophy of science at that time (Popper, 1959, 1963), the early writings of such influential philosophers as Feyerabend (Feyerabend, 1964; Feyerabend, 1962), Habermas (Habermas, 1963) and Foucault (Foucault, 1966; Foucault, 1973; Foucault and Rabinow, 1997), were emerging in philosophy more generally. Certainly the early 1960s represented a period of significant change, and the history and sociology of science both enjoyed burgeoning popularity from that point to the present day.

In terms of integration *The structure of scientific revolutions* represented a significant blow to the façade of the academies that science was comprised of the steady accumulation of immutable truths. Despite concurrent and later criticisms (Kuhn, 1977; Fuller, 2000; Sardar, 2000), Kuhn convincingly demonstrated that science was a process of truth-seeking, not a storehouse of absolute truths. Further, what passes for scientific truth is the product of a range of human decisions, not all of which are scientific. Taking an historical perspective, Kuhn drew science into the realm of human activity, in all its uncertainty and fallibility. As such it became comparable to *other* human systems of activity.

While Kuhn's writing, supported by others of that period, may have opened the doors to view science in its pyjamas, so to speak, it did not in itself pose a major challenge to the authority of science. While Kuhn described the validity of science as partial and contingent, his analysis did not directly assert that this was problematic, or challenge whether it was still the best there was. It was left to the sociology of science, in conjunction with post-empiricist philosophers to take that next step.

Sociology of science, and the sociology of scientific knowledge The sociology of science as it is known today has an identifiable history that extends back to 1938 with Robert K. Merton's paper "Science and the social order" published in *Philosophy of Science* (Merton, 1938). This work, and others by Merton throughout the 1940s and 1950s established the view that science can be regarded as a suite of socially-defined norms and imperatives. Thus science was argued to be a social activity as well as a technical–empirical one, where the social organisation was functional, in that it served to sift the good science from the bad, and reward only the good science.

In the mid-1970s, this functionalist view of the sociology of science came under attack from the so-called 'Strong Programme', based at the University of Edinburgh, primarily under the guidance of Barry Barnes and David Bloor (Barnes, 1977; Barnes and Bloor, 1981). The Strong Programme drew on language-based philosophies to assert that science does not rely on 'direct' experience of the real world as its basis, but rather is an 'interpretive' activity, in which scientists impose their own interpretive structures on the things they see. It draws on there being a 'looseness of fit' between the language and the world (Kim, 1994), and that breaching that looseness of fit is an act of interpretation and persuasion. These interpretations, then, are not based on 'fact' or 'observation', but on a system of scientific *conventions*.

Barnes and Bloor and the Strong Program represents a significant shift in the sociology of science in their assertion that the development of scientific knowledge is essentially a social process, *not* a biophysical one. This perspective, more so than the historical or functionalist schools of thought, contributed to the democratisation of knowledge by highlighting and challenging the hegemonic nature of conventional scientific structures and practices. Scientific conventions, with no direct link to the physical world, may therefore be comparable with other conventions found elsewhere in other socio-political structures.

Science studies While science studies (or science and technology studies) is closely allied with the sociology of science and sociology of scientific knowledge, and the three schools interact, science studies focuses on ethnographic, anthropological-style investigations of scientific practice. Following the seminal ethnographic work of Latour and Woolgar (1979), science studies emerged as a distinct genre of research into the practice of science.

Science studies have included extensive anthropological-style studies of work in laboratories and in the field, (Latour and Woolgar, 1979; Knorr-Cetina and Mulkay, 1983; Latour, 1987; Knorr-Cetina, 1999), as well as cultural studies of the role of science and technology in culture, and *vice versa* (Nelkin and Lindee, 1995), and interactions between science and policy (Jasanoff, 1990). Science studies authors are predominantly committed to social constructivist or constructionist epistemologies, and tend to focus on the ways in which science, society and technologies are co-constructed. By characterising science as a construct, science studies has helped to build the theoretical foundations of the knowledge democracy, and added a slightly different dimension to the 'social' challenge to the traditional concepts of objective, logical, rational science.

In summary, the various arms of 'social' studies of science have complemented the ontological and epistemological shifts noted in participatory research, away from the popular conception of science as representing 'real', 'legitimate' knowledge towards science being one of many different ways of coming to know the world. They have generally been associated with concerns to address the balance of power between what has been perceived as the hegemonic power of the scientific institution and the disempowered lay citizenry. However, while the validity of these concerns is not to be denied or downplayed, the emphasis of the knowledge democracy on integrating the needs and concerns of those outside academia does carry risks that are not well documented in the knowledge democracy literature. The commonly recognised risk is that a policy of engagement diminishes the independence and thence the credibility of research. Less commonly recognised is that in order to influence bodies that are *already* powerful (government, industry), science needs to maintain its (hegemonic) authority. If the credibility needed by researchers to be taken seriously as independent assessors of powerful institutions is reduced, an important mechanism for social critique may be lost.

THE KNOWLEDGE AUTOCRACY

This latter concern has supported forces resisting such change from the knowledge democracy, labelled here the 'knowledge autocracy'. The knowledge autocracy is concerned with the defence of traditional understandings of science itself—how it is performed, how it comes to generate 'truth' or 'facts', and how it gains influence in the world. Traditional concepts of science as the major source of reliable, authoritative knowledge about the world retain a strong hold on the scientific self-image and scientific practice. Yet the ideology of the knowledge autocracy is already a battlefield, regardless of knowledge democracy (or knowledge economy) arguments for integration.

This battle is taking place between the traditional 'autocratic' ideology of science and the social studies of science, discussed at the end of the previous section. While not directly concerned with integration, the defence of the autocracy against social critiques is an important part of the fractured landscape of integrated research, as it centres on the values that underpin traditional understandings of how science comes to take meaning and influence in the world. By attempting to overcome the structures that science has relied upon to support its traditional position, it is easy for appeals for integrated research to invoke the dilemmas and battles of the knowledge autocracy and the 'Science Wars'.

The autocracy strikes back

In general, the public and academic questioning of science in the post-World War II era noted in Chapter 1 supported the social critiques of the privileged, 'objective' status claimed by scientists. The backlash against these critiques has generated ongoing, often acrimonious exchanges between representatives of different camps, which have earned the title 'the Science Wars.'

Since the mid-1990s the Science Wars have burgeoned from small-scale skirmishes to prominent battles. The essence of the Science Wars is the belief of 'conventional' scientists that, by highlighting the socially- and politically-constructed nature of research, social studies of science have sought to undermine the credibility of science. The conventional scientists reacted with claims that science studies was founded on 'unscientific' bases. For example, in 1996, in the preface to an edited volume of the Annals of the New York Academy of the Sciences entitled *The flight from science and reason*, Paul Gross wrote:

> We believe that there is today in the West, among professors and others who are paid, in principle, to think and teach, a new and more systemic flight from science and reason. It is given endless and contradictory justifications, but its imperialism—for example under the banner of "science studies"—and the highest esteem in which it holds the trendiest irrationalisms, are undeniable. (Gross et al., 1996, p. 2)

The same year, physicist Alan Sokal authored a now (in)famous hoax paper, published in the sociological journal *Social Text* (Sokal, 1996). This paper, and the ensuing debate over its significance (or lack thereof) sparked a parley between conventional science and various arms of the sociology of science and science studies (for summaries see Gieryn, 1999; Sardar, 2000), which continues to the present day (see, for example, Ashman and Baringer, 2001; Brown, 2001; Weinberg, 2001).

The implications of these arguments for integration are ambiguous. On the one hand, integration across disciplines may be tolerated, even supported, as an appropriate way for scientists to make research outcomes more relevant to policy-makers and others. However, the direct interaction and engagement of scientists and non-scientists, by denying the possibilities of independence and objectivity, constitutes a 'democratic' threat to the integrity of science. Consequently, attempts to practice integrated research can be readily criticised from the perspective of the knowledge autocracy, as it is seen as allowing—or even encouraging—external factors (such as politics) to compromise the credibility of the institution of science. Such criticisms highlight the risks of integrated approaches that seek to engage beyond the academy, as they have important implications for the ability of researchers to independently analyse and be critical of the policies and practices of other institutions. This tension between traditional ideas of science and integrated research effectively represents two different ways in which science can bring about change (by engagement or by disinterestedness), and is an important feature of the fractured landscape of integrated research.

The ambiguity of the knowledge autocracy in policy

The ambiguity of the knowledge autocracy with respect to the need for science to take a more active role in the decision-making and action-taking worlds is also apparent in science policy. Some international illustrations include the United Nations Declaration on Science, the Kyoto Protocol, and a national example is a recent report by the Australian Chief Scientist.

United Nations Educational, Scientific and Cultural Organisation In 1999 members of the United Nations, through the United Nations Educational, Scientific and Cultural Organisation (UNESCO), agreed to the *Declaration on science and the use of scientific knowledge*, which centred on commitment to the document *Science agenda— framework for action* (UNESCO, 1999). The importance of scientific work across disciplinary boundaries in the context of environmental change was emphasised:

> Interdisciplinary research involving both the natural and the social sciences must be vigorously enhanced by all major actors concerned, including the private sector, to address the human dimension of global environmental change… and to improve understanding of sustainability as conditioned by natural systems. (UNESCO, 1999, Section 31).

While interdisciplinary integration is clearly endorsed, how interdisciplinary research is to be 'enhanced' by major actors is unclear. To further confuse the issue, the Science Agenda does not differentiate between research geared towards production and research geared towards social change. The introduction to the framework summed up this duplex view of science, vaguely stating that "Science policy should promote the incorporation of knowledge into social and productive activities" (UNESCO, 1999, Section 38).

Consequently this statement has opted for interdisciplinary research as a relatively 'safe bet' in balancing the need for integration against the traditional ideals of science.

The Kyoto Protocol

In 1993 the Kyoto Protocol brought scientific research and the scientific community well within the political arena. This went beyond the usual 'expert advisory' provision of technical information about climate change, as countries' adherence to the protocol is to be judged by a scientific panel:

> each Party included in Annex I shall provide, for consideration by the Subsidiary Body for Scientific and Technological Advice, data to establish its level of carbon stocks in 1990 and to enable an estimate to be made of its changes in carbon stocks in subsequent years. (UNFCCC, n.d., Article 3.4)

This suggests that, in terms of integration, the Kyoto Protocol has granted the scientific community a significant role in the political/legal system beyond the provision of advice towards assessing nation-states' compliance. While this led to the close entanglement of science and policy, it was ambiguous with respect to the knowledge democracy/autocracy. To the extent that science was being brought into the democratic systems of the United Nations, it can be regarded as a step towards a knowledge democracy, a recognition that politics alone cannot solve the problems posed by climate change. On the other hand, the Kyoto Protocol can also be read as simply increasing the hegemony of the already-powerful (the scientists and the political negotiators) by bringing them together, in turn decreasing the ability of those 'outside' to have a voice.

THE KNOWLEDGE ECONOMY

The third strand of literature relevant to the fractured landscape of integrated environmental research is the 'knowledge economy'. The knowledge economy can be summarised as a widespread perception that the acquisition and transformation of knowledge has replaced the acquisition and transformation of raw materials as the engine of economic growth (OECD, 1996). As such, the primary mechanism for affluent 'new economy' nations to continue increasing their wealth is knowledge and information. In terms of integration, the knowledge economy is concerned with the connections between research and the production of goods and services.

In a similar manner to the changes described as the democratisation of knowledge, the understanding of the relationship between research and development and the production of goods and services has also undergone a significant sea-change over the last decade. However, to document the change, some description of the previous situation is necessary.

From R&D management to innovation systems

Research and development (R&D) management concepts overlapped with agricultural R&D concepts, although the two literatures tended to develop in parallel without much connection. Indeed, the concept of 'transfer of technology' as is commonly applied in commercial, technological sectors today was in large part derived from its application in agriculture (Vanclay and Lawrence, 1995). The commercial/industrial version of the transfer of technology model is illustrated in Figure 5.

Figure 5. The 'technology push' model of the innovation process.

Yet in a similar process to that described earlier in this Chapter, the simplicity of the transfer of technology model has also been challenged in the industrial sector. In the late 1980s and early 1990s, several authors proposed that the linear models of technology transfer were insufficient as descriptions of how innovations were transformed into economic growth (Freeman, 1987; Lundvall, 1992; OECD, 1996). Perhaps unsurprisingly, the business sector's criticisms of these narrow, linear models were remarkably similar to those of the agricultural sector. Fields such as organisational learning (Argyris and Schön, 1996) emphasised the complexity of innovation *within* firms. The field of innovation studies pointed to the extent that research was integral to a much broader, more complex system *beyond* firms.

Innovation studies   Innovation and its role in generating profit is not a new topic; it has been a core part of R&D Management since its inception (see, for example, Burns and Stalker, 1961; Byatt and Cohen, 1969; Taylor, 1971). However, this early literature typically focused on innovation within firms. More recently, Freeman (1987) introduced the concept of 'national systems of innovation', a theme that was taken up by Lundvall (1992) in a highly influential volume that shifted the focus of innovation from individual firms to a larger industry–nation-state interaction. Lundvall introduced the concept thus:

> a system of innovation is constituted by elements and relationships which interact in the production, diffusion and use of new, and economically useful, knowledge… [A] national system encompasses elements and relationships, either located within or rooted inside the borders of a nation-state. (p. 1)

In other words, it focused on the interactions between the nation-state and industry as the site of innovation that led to the general growth and prosperity of the nation as a whole. But it also contributed to the increasing recognition of the role of science and its connection with society, especially industry, as a vital component of that system. As Quinn et al. (1997) write:

> As a society, and as managers, we need to understand the truly chaotic way in which science, innovation, and technology develop and learn how to manage it in this chaos rather than attempt to convert the process into something it will never be: orderly and predictable. (p. 19)

In other words, there was a shift in thinking away from the linear models to ones that emphasised the interconnectedness of research within broader systems of commercial production.

Enter the knowledge economy

These shifts in thinking supported a new perspective on both research, and the concepts of society and economy more broadly. The boom of the information and communication technologies industries in the United States and Europe in the late 1980s and early 1990s spectacularly confirmed the centrality of research and technological development in economic prosperity. The knowledge economy was born.

The impact of this sea-change in economic thought on research, both public and private, is perhaps yet to be fully realised and is not yet well understood. But at a general level the knowledge economy has placed research, its development and application at the forefront of economic, governmental and organisational planning.

This has not simply been a matter of rearranging the components so that knowledge 'producers' are granted higher status. As the language suggests, the knowledge economy frames the knowledge produced in all sectors of the economy as a *commodity* with current or potential commercial value. Consequently a significant manifestation of the knowledge economy has been broad conceptual and practical shifts within research institutions, not only privately funded ones, away from scientific knowledge as public good towards its privatisation. In the knowledge economy, the protection of intellectual property is paramount, because it forms the basis of commercialisation potential.

Consequently, the dominant ways of thinking about research and the publicly funded institutions whose primary function was to generate 'new knowledge' underwent a fundamental change. In combination with the theories of innovation systems, the knowledge economy placed these institutions closer to industry, with the emphasis no longer on 'feeding into' the private sector, but actively 'working with' them. This shift was noted and endorsed by the Organization for Economic Co-operation and Development (OECD) in 1996, when they wrote that:

> The science system, essentially public research laboratories and institutes of higher education, carries out key functions in the knowledge-based economy, including knowledge production, transmission and transfer. But the OECD science system is facing the challenge of reconciling its traditional functions of producing new knowledge through basic research and [education] …with its newer role of collaborating with industry in the transfer of knowledge and technology. (OECD, 1996, p. 7)

As such, the knowledge economy focused highly influential and broad-reaching attention on overcoming the barriers to integration among researchers and between researchers and those arms of the economy that convert research into wealth.

The consequences for publicly funded research have been ambiguous, perhaps even contradictory. While it is widely recognised that government support for 'basic' research is necessary as the foundation for applied research (although this simple distinction is itself becoming less clear), there is simultaneous pressure for more research to be focused on meeting production-based outcomes. Research is at once more important, but also more accountable and more open to questioning from non-scientists, as their knowledge is recognised as valuable too. In this way, some of the implications of the knowledge economy for research practice are superficially similar to those of the knowledge democracy.

The emergence of the knowledge economy was significant to the development of the concept of integration in several ways. In one sense it reaffirmed the idea that it was not research *in isolation* that added to a nation's wealth-growing capacity, but research *in connection* with other economic bodies, in particular business and industry. Following from the popularity of organisational learning, in the early to mid-1990s the new term of 'knowledge management' became the latest managerial juggernaut. Knowledge management reflected the key reorientations of the knowledge economy away from thinking of production as being solely (or primarily) concerned with the transformation and flow of goods and services to being concerned with the transformation and flow of knowledge.

The knowledge economy generated several shifts in managerial thinking, including the recent buzz of 'knowledge management' (for an overview see Liebowitz, 1999) and 'alliance management' (Doz and Hamel, 1998; Harbison and Pikar, 1998). While these are not central to the history being recounted here, they are significant to the extent that they have opened up many new opportunities for managers and researchers to experiment with new organisational relationships.

In some respects these were similar developments to the democratisation of knowledge, to the extent that they can be attributed to the same growing appreciation of the complexity and non-linearity of relationships between science and society. However, the epistemological and moral motivations of the knowledge economy and the knowledge democracy are in stark contrast. The focus of innovation studies was on the more efficient transformation of knowledge into saleable product, hence it was grounded in a neo-classical economic perspective, in which the primary aim of research is to generate economic wealth. Morally, then, the knowledge economy is grounded in the utilitarian view that the pursuit of individual self-interest brings the greatest benefit to society as a whole. Epistemologically, the knowledge economy concurs with positivist views that science sits at the foundation of much innovation, and as such maintains science as privileged access to knowledge about the biophysical world.

The knowledge economy research models

The impact of the emergence of the idea of the knowledge economy has been observed and, to an extent, theorised by several researchers, with popular models being "Mode-2" research and the "triple helix".

The New Production of Knowledge: Mode-2 research Mode-2 research has become something of a catch-cry in research policy and management circles, attesting to the timeliness of the publication of *The new production of knowledge: the dynamics of science and research in contemporary societies* by Gibbons et al. in 1994, followed up by a similar team in 2001 (Nowotny et al., 2001). Gibbons' theory is based on the observation of newly (or recently) emerging research structures (Mode-2) that are different from conventional scientific structures of knowledge production and dissemination (Mode-1). The aim of their analysis is to "clarify the similarities and differences between the attributes of each [mode], and help us understand and explain trends that can be observed in all modern societies" (p. 1).

The authors' starting point is the claim that knowledge, which was formerly produced in academic institutions, particularly universities but also government research organisations and corporate laboratories, is now being produced by a much wider range of actors, in contexts that are more social and economic than academic. The divisions between academia and other institutions are being blurred in the attempt to produce innovative products (in the private sector) and innovative solutions to social and environmental problems (in the public sector). As such, they are claiming that similar processes are at work in both—that continuing fragmentation and specialisation of knowledge has led to a greater emphasis on developing formal and informal links not only between different disciplines, but different types of organisations altogether.

The ability to develop and maintain many different types of relationships with a wide range of both scientific and non-scientific players is key to successful Mode-2 research. They also place particular emphasis on change in scientific institutions, whose traditional ways of producing and disseminating knowledge have become inadequate in the constant search for competitive advantage in the more globalised economy. In Mode-2 research, the ways in which the research is going to be used frames the research in the first place—problems to be tackled are not intellectual problems dictated by scientific curiosity or the progress of the discipline or sub-discipline, but are generated by the need to address a particular 'real-world' issue. As the authors acknowledge (p. 4) this is not a particularly new type of research: the so-called 'applied sciences' such as engineering or medicine have always been more Mode-2 than Mode-1. More importantly though, the context of application in Mode-2 research is not decided within the academic domain alone, but in negotiation with various other actors. Hence it is not the engineers or medical doctors who decide which research is most pressing, but a broader range of actors or stakeholders: "...it is shaped by a more diverse set of intellectual and social demands than was the case in many applied sciences...' (p. 4).

While the authors refer to environmental issues and research as exemplifying Mode-2 research, they generally assume that the same principles of Mode-2 research apply regardless of public or private good status. Indeed, part of their justification of the growing prevalence of Mode-2 research is that the distinctions between public research institutions and private organisations are becoming blurred, as collaboration between the public and private sectors increases. The 'social contract' that brings about the closer ties between research and non-research sectors and makes private industry more accountable to government and citizens is presumed to be the same in, for example, pharmaceutical research as it is in climate change research.

It is acknowledged that the boundaries are no longer clear-cut (if, indeed, they ever were), especially due to the emergence and strength of 'environmental' industries and consumer lobby groups. However, Gibbons et al. don't tend to separate out the different forces that have been pushing towards greater integration, and hence the potential for conflict between the knowledge democracy and the knowledge economy is understated. Lack of historical contextualisation in their work allows the authors to observe the superficial similarities between the push for greater integration in public and private good research and presume they are the same. As this chapter has tried to illustrate, the differences between the two can be readily observed from an historical viewpoint. The analysis presented in this Chapter suggests that the moral and epistemological differences between the knowledge economy and the knowledge democracy are likely to be significant.

Triple helix   The triple helix model was proposed in the mid-1990s by Etzkowitz and Leydesdorff (1997). In its simplest form, the triple helix simply refers to any configurations of industry, government and science working together in an integrated way (Ernø-Kjølhede et al., 2001). The helix metaphor suggests the entwining of these three strands into a shared or common entity. As such, it focuses attention on the role of science policy in shaping relations between these groups, or configurations of the triple helix.

Etzkowitz and Leydesdorff (2000) describe three types of triple helix relations: in version 1, the nation-state controls both industry and science, and dictates the relationships between them (e.g. Former Soviet Union and Eastern European countries); in version 2, the three institutional sectors are strongly divided and relations across them are highly circumscribed (e.g. some aspects of Swedish and US science policy); and finally triple helix version 3, where institutional spheres overlap with "each taking the role of the other and with hybrid organizations emerging at the interfaces" (p. 111). They illustrate this version as Figure 6.

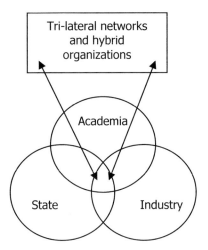

Figure 6. The triple helix model of university–industry–government relations.
(Source: Etzkowitz and Leydesdorff, 2000, p. 111)

The triple helix model represents a significant shift from the linear transfer of technology model to an appreciation of the interactions between science, and business and government as a complex, non-linear, inherently unstable co-evolution between three dynamic systems (Etzkowitz and Leydesdorff, 2000).

Increasingly science is viewed not merely as a source of informational input to these systems, but as an active (and reactive) participant in an ongoing process of transition and renewal.

Once again, the triple helix model tends to treat research as a homogenous whole that is focused on research that has application in its transformation into private goods. Yet some of the assumptions that can be made at this general level of science–government–industry interaction do not apply in the case of public good research, such as profit being the driving force behind the increasing interaction. While the profit motive may have legitimacy in some environmental research sectors, especially as market mechanisms are turned to as a way of internalising external environmental costs, as illustrated earlier in Figure 2, there is a range of influences driving towards new research models. Similarly, while the role of governments as regulators of environmental issues renders them more prominent in the environmental 'triple helix' configuration than in profit-driven sectors, other sectors, such as 'community' are excluded. Other models, or variations to the triple helix and Mode-2 models are needed to explain the extension of research transformation beyond the profit-based, private-good research base, into 'public good' sectors such as environmental research.

Integration and the knowledge economy in policy

As for the knowledge democracy, the ideology of the knowledge economy is also supporting calls for integration in policy. While the international support of the OCED for the knowledge economy has already been noted, recent examples in Australia include a major report by the Chief Scientist and the Commonwealth Scientific and Industrial Research Organisation's (CSIRO's) Strategic Plan.

Chief Scientist's report   In Australia, a recent review of science policy conducted by the Chief Scientist (the most senior science policy adviser to the Federal Government) also emphasised the importance of integration with respect to the knowledge economy. He reported that:

> Integrating the innovation system across all points can increase the chance of generating more products and processes that enhance our lifestyle. The innovation system is dependent on strong links between all players, government, industry and research performers. (Batterham, 2000, p. 11)

and further:

> By and large, our competitors and economic partners are adopting different combinations of integrated measures to strengthen their capacity to innovate. Although the pace of progress across these countries fluctuates, they are constant in their drive towards knowledge-based economies. (Batterham, 2000, p. 41)

It is the knowledge economy, and the place of science as a key driver in economic growth and productivity that dominates this review. The overtones of this report are that the research community has a moral obligation to commercialise their research and engage in the knowledge economy.

CSIRO strategic planning   The Commonwealth Scientific and Industrial Research Organisation (CSIRO) is Australia's largest research body, and one of the largest and most diverse scientific research organisations in the world. It carries out research across a wide range of sectors, including agricultural and environmental research. CSIRO is highly respected, both nationally and internationally. At the level of strategic planning, CSIRO's goals and aspirations illustrate the ambiguous position of public research organisations within a knowledge economy/democracy. In their 2000-2003 Plan, they state:

> We have decided that many of the land and water issues affecting the sustainability of agriculture, mining, mineral processing, manufacturing and the built environment can best be dealt with via large-scale integrated work … with appropriate advisory and management mechanisms in place to ensure that customer groups in the relevant production-based sectors can exert appropriate influence on those projects. The objective is that work will be undertaken so as to meet the sustainability needs of customers in production sectors while addressing the environmental issues through a systems approach to broad-scale land and water management. (CSIRO, 2000, p. 8)

In other words, the integration of science is supported, as is integration with non-scientists—significantly called "customers". The implication of the term is that the knowledge economy is the dominating force: they will be paying for access to CSIRO knowledge, which is a commodity for purchase. Further, it implies that those *not* paying are excluded from the integrative context.

In CSIRO's environment sector, their capacity to do 'big science', integrated across discipline and industry sectors is promoted:

> CSIRO's environmental activity… derives from a disciplinary base across all elements of the environment, a breadth rarely found in one institution, anywhere. Increasingly, this knowledge is tested and applied in big, integrative, collaborative activities, focused on specific contexts, regions and problems, and delivered 'on-the-ground' in conjunction with a range of partners and co-investors. These integrated responses take account of economic, social and institutional factors and draw on relevant skills in CSIRO and collaborators. (p. 13)

The nature of the relationship between CSIRO and "partners" is unclear, but the description of "co-investors" once again strongly invokes the commodification of the research rather than its democratisation.

Effects on public good research?

As these models and policy statements have indicated, the effects of the knowledge economy on research that is concerned with 'public good' problems, such as environmental research, are ambiguous. Despite most of the knowledge economy models bundling public and private good research together, as yet none has adequately examined how or why this should be the case. The fundamental distinction between public and private goods has not changed. Not only is much environmental research 'basic' in the scientific sense of needing to investigate fundamental processes that are conceptually a long way from any technical application on the ground, but even that on-the-ground application itself is often not productive in the sense of directly generating income. More commonly it incurs a cost that may be highly cost-effective in terms of avoiding larger costs in the future, but these benefits are themselves uncertain. As such, environmental research—and environmental restoration or protection—remains a public good.

Despite this, there has been a strong move within Australia towards a 'purchaser–provider' model as the key link between government and environmental research. While this will be discussed further in later chapters, the purchaser–provider model draws on the concept of the knowledge economy, and characterises environmental knowledge as a commodity that can be transferred (without significant cost) from producer to consumer. This has facilitated the privatisation of public good research through the creation of artificial markets where environmental researchers compete against each other for the provision of the said knowledge commodity.

While this does not affect the public good status of environmental actions taken on the basis of the knowledge provided, it does shift environmental research onto a different moral and ethical platform, which has significant practical ramifications. In particular, intellectual property provisions demand that researchers do not share their research results if doing so will help them maintain an advantage in the next funding round *except* with their contracted research purchaser. The contractual purchaser–provider model (or variations of it) can be construed as integrative, in the sense that it evidences a strong relationship between science and a research 'user'. This is in stark contrast to the knowledge democracy, which demands that many different players should contribute to the construction of knowledge, and that decision-making should be an open, collaborative process. The concept of 'integration' may potentially encompass both democratic and economic motivations for breaking down traditional scientific barriers.

INTEGRATION: DEMOCRATISATION AND COMMODIFICATION

So far this chapter has attempted to show that integrated environmental research is in the unusual position of being a product of both the democratisation *and* the commodification of knowledge. These processes have developed and strengthened simultaneously, with similar superficial impacts—calls for closer ties between science and non-science institutions, focus on science in the context of its application, greater accountability, demand for more interdisciplinary or trans-isciplinary work, and an emphasis on innovative approaches. Yet beneath these superficial manifestations lies the potential for deeper conflicts that have yet to be fully analysed.

These closer ties are ambiguous with respect to the knowledge autocracy. While the knowledge autocracy resists democratisation of knowledge, its position in relation to commodification is less clear. In one sense, it is rejected, as contract-based research diminishes the researcher's autonomy, which is an integral part of the autocracy. Yet the knowledge economy also reaffirms the hegemony of scientific knowledge, and emphasises its monetary value, as well as its value in application more generally. This is appealing to many researchers, especially as it offers new avenues for recognition, including financial reward, in times of falling block grants, non-tenured positions, and increasing competition for the research funding dollar.

Interconnections

These diverse literatures have all contributed in different but connected ways to the concept of integrated research. Major developments in sociology and philosophy, together with widespread realisation that environmental issues are not only scientific but need to be balanced with non-scientific concerns and expertise, have created a groundswell labelled here 'the knowledge democracy'. The knowledge democracy has been implemented in environmental research through a wide range of community-based participatory research approaches. Broadly speaking the knowledge democracy is based on widening the concept of legitimate expertise to include non-scientists—from community groups to non-government organisations, to government and industry—in environmental decision-making.

These changes are being resisted by the knowledge autocracy, those who continue to endorse the sanctity of science as a superior, rational knowledge that deserves a special place in the decision-making world.

In contrast, developments in economic thinking have driven a view of knowledge as a commodity. As such scientific knowledge needs to be protected by intellectual property rights and integrated with the productive sectors through commercial exploitation. This perspective reifies scientific knowledge in economic terms, and encourages purchaser–provider relationships between science and non-science sectors.

These are each evident in science policy and environmental policy at national and international scales to varying degrees. Consequently, environmental research sits amidst these strangely contrasting, converging, potentially conflicting ideologies and criteria for what constitutes good integrated environmental research, as illustrated in Figure 7.

The tension illustrated in the centrepiece of Figure 7 is not only the result of ideological clashes between the knowledge economy, autocracy and democracy, but are also due to policy clashes between science policies and environment

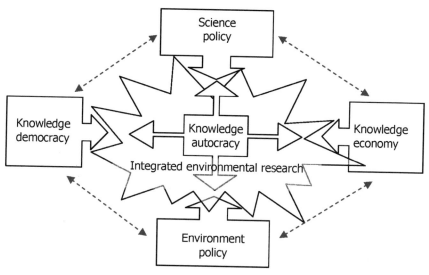

Figure 7. Forces of integrated environmental research.

policies that encourage and reward different conceptions of integration. Further, 'internal' tensions with conventional knowledge autocracy views of science suggest integrated environmental research is 'spilling over' the central, autocratic core of the institution of science, again leading to tension and uncertainty.

These different forces both act directly on the conduct of integrated research (when, for example, researchers make a commitment to a participatory research approach), and influence each other indirectly, leading to unpredictable interactions, tensions and confusions (for example, working 'in partnership' with one's 'customers').

The overall aim of this chapter was to offer an explanation for the complexity, ambiguity, and confusion that forms the landscape of integrated research. The tensions that can result are not only due to the potential for conflicting 'external' requirements imposed by others but also because they often reflect personal, moral commitments. As such the contrasting epistemologies that underlie the knowledge democracy, the knowledge autocracy and the knowledge economy are both abstract philosophical issues and practical dilemmas researchers deal with in their own understandings and applications of the concept of integrated research.

Consequently, integrated research takes place not only in the more commonly cited context of fragmented disciplines, organisations and institutions, but also in a landscape of ethical, practical, political and scientific dilemmas. This landscape reinforces the significance of the questions at the heart of this study: how is integrated environmental research being done, amidst this confusion? What lessons can be learned from those experiences that can inform the theoretical and practical development of this type of research?

CHAPTER 3

CASE STUDIES—THE CONTEXT

As discussed in Chapter 1, one of the mainstays of this study's methodology is the notion of *social communicative practice* as a process of ongoing negotiation with respect to historical and contemporary context. The background of the cases involved in this study is vital to the remainder of the study, not in the passive sense of a backdrop, but as a significant part of the resources people drew on in developing and expressing their understandings of integrated environmental research. This chapter will outline several key dimensions of context that have a bearing both on why these organisations were chosen as cases, and on how the conversations unfolded.

Of course, my own 'context' plays a key role here too. In particular, my research aims, as given at the end of Chapter 1, dictated that the study be based around situations where some version of the concept of integrated environmental research was being implemented. Preferably, it would be within a context that inhabited the somewhat vexed environmental research landscape, positioned between the knowledge economy and the knowledge democracy, described in Chapter 2.

These main features were met by Australian Cooperative Research Centres.

COOPERATIVE RESEARCH CENTRES

Cooperative Research Centres (CRCs) are natural science and engineering research organisations formed by formal agreements between extant organisations and the Australian Federal Government. They are 'virtual' centres, in the sense that they do not physically share the same location. Instead, they capitalise on the resources that already exist but are dispersed across different organisations, by funding researchers to work cooperatively with others involved in similar problem- or issue-oriented fields, but different institutional contexts.

The CRCs are unique in several ways. The partner organisations that make up a CRC may comprise a wide range of research and non-research organisations and firms; their organisational structure insists on a degree of cooperation and integration across those groups; and they have stringent accountability and reporting requirements. While other similar arrangements are emerging in other parts of the world, perhaps the most unique feature of the CRC Program is their relatively long history. Having commenced in 1991, the program was a number of years 'before its time', in terms of the pressures towards greater integration noted in Chapter 2.

At the time this study was conducted, the overall objectives of the CRC Program were described by the Federal Government as:

To enhance the contribution of long-term scientific and technological research and innovation to Australia's sustainable economic and social development;

To enhance the transfer of research outputs into commercial or other outcomes of economic, environmental or social benefit to Australia;

To enhance the value to Australia of graduate researchers; and

To enhance collaboration among researchers, between researchers and industry or other users, and to improve efficiency in the use of intellectual and other research resources. (CRC Program, 1999, p. 1)

Mercer and Stocker (1998) describe the CRC program as "a bridging mechanism linking the public sector research and higher education organisations and the users of new knowledge, from the private and public sector."

CRCs are co-funded by the Australian Federal Government and their partner organisations, with government funding allocated on a competitive bid basis. They have an initial lifespan of seven years, with an option to bid for renewal at the end of that period. There are usually between 55 and 65 Centres in operation at any given time, ranging across six sectors: environment; manufacturing technology; information and communication technology; mining and energy; agriculture and rural based manufacturing; and medical science and technology.

The historical context

The CRC concept was developed and championed in the late 1980s by the then Chief Scientist to the Federal Government, Professor Ralph Slatyer, with the first CRCs established in 1991. The program developed in response to the perception by the Federal Government that there were weaknesses in the national innovation system, including:

- disincentives to collaborate among researchers and business;
- weak links between organisations and users;
- lack of critical mass due to the institutional and geographical dispersion of Australian research and application;
- lack of mobility of key personnel; and
- difficulty and expense of maintaining international links in an isolated country. (Mercer and Stocker, 1998).

In a first-hand account of the origins of the CRC Program, Professor Slatyer wrote:

What I envisaged to address these weaknesses was… a 'One Stop Shop' for innovation, consisting of a cooperative team of researchers and research users, drawn from various organisations, and of adequate size and composition to have a real and continuing impact in the sector where it was located. (2000, p. 10)

In addition, there has been a long-standing perception within the community that many of Australia's best intellectual resources do not remain in Australia— whether as talented individuals who move overseas to further their careers, or as patents that are developed for commercial benefit elsewhere. While there is some debate as to whether this perception is substantiated (Gaylord, 2000), the 'brain drain' is sufficiently embedded in the Australian cultural psyche for it to be an issue of concern for government. Further, in the context of the knowledge economy, the development of the CRC Program was also perhaps a response to a perception that the Australian public were not receiving the levels of economic return for public funds invested in research that they should (or could). In a general sense CRCs were instituted to counter these weaknesses, by providing opportunities for scientists to develop close ties with industry, facilitating commercialisation 'at home'. A key aim of the Centres from the outset, then, was to enhance integration across many scientific boundaries.

Over the ten years since their initiation, CRCs have gained a prominent position in Australian research. In terms of funding, the total Federal Government funds committed since 1991 and running through until 2006 are AU$1.5 billion (about AU$140 million a year, with each Centre receiving on average about AU$2.2 million per year). Over the same period, CRC core partners have committed about AU$4.0 billion to CRCs (CRC Association, 2000). They have been cited regularly as exemplars of industry—government—science collaboration, and in the 2001 Australian Federal election received endorsement and pledges of further support from both major contesting political parties. The Chief Scientist's report of 2001, mentioned in Chapter 2, also gave favourable mention to the CRC Program.

Despite their relatively small budgets, the large number of research organisations involved in CRCs means they were well-known within the Australian science and research sectors. Although there were no figures available regarding the proportion of Australia's science community who are involved with one or more CRCs, anecdotally it was quite common for researchers in the environment sector to either be part of a CRC now, or to have worked with one in the past. Their prevalence ensured that even if researchers had not directly worked within a CRC, they often had colleagues who had, and hence know of the CRC Program.

The main emphasis of the CRC Program has been in the domain of the 'knowledge economy', although the term was in its infancy when the Program was established. Nevertheless, public good issues, such as the environment, agriculture and public health have also formed a major part of the CRC structure, comprising three of the six CRC research sectors, and 39 of the 65 Centres that were operational in 2000. This does not mean that over half the CRCs are oriented to public good issues though; within each of those sectors were several CRCs just as devoted to commercialisation as those in manufacturing and information and communication technology. Yet, especially within the environment sector, and to a lesser extent the agricultural and health sectors, public good issues remained prominent.

However, the need to involve 'end users' that drove the industrial sector CRCs was also a driving force within the public good CRCs. As noted in Chapter 2, the action required at that broad level of 'involve end users' was applicable across public and private good research contexts. The Mercer Review of the CRC Program (Mercer and Stocker, 1998) noted the disadvantages facing the public interest CRCs when being compared against organisations with strong industry support and commercialisation activities. This was especially significant given their Terms of Reference to consider ways of improving commercialisation and self-funding. The authors wrote:

> One in five of the CRCs are focussed on environmental management, and along with several CRCs in the medical field, are concerned primarily with public interest outputs. Forty percent of CRCs are directly related to the primary industries and often aimed at improvements in productivity and sustainability among large numbers of dispersed users. In such cases the wide dissemination of new knowledge, rather than its appropriation by a few firms, is vital if it is to have a significant commercial impact. (Mercer and Stocker, 1998, p. v)

While the reviews to date acknowledge that some of the criteria for judging the public sector CRCs are necessarily different from those driven by commercialisation, CRCs in all sectors continue to rationalise their 'value adding' in monetary terms. The theme of the 2001 CRC Association conference, "Return on investment", highlighted the significant and somewhat creative lengths public good CRCs went to in order to place dollar values on their outcomes. While some of these were presented with tongue slightly in cheek, given that CRCs are assessed competitively not only when they are funded but also when they reapply for funding at the end of their seven-year contract, there is a pervading sense that the public good CRCs are competing on an uneven playing field.

Having said that, there was little direct evidence that the public good CRCs were disadvantaged as a result of this imbalance. In 2000, for example, there were more CRCs in the environment sector than in any other sector. This suggests that the selection committees were flexible in their approach to criteria of commercialisation, but this did not negate the formal requirements. In terms of this study then, the 'public good' environment sector CRCs were likely candidates for experiencing some degree of tension between the various forces of integration discussed in Chapter 2.

The organisational context

In terms of the CRC Program as a whole, the Program was an arm of the Australian Federal Government's Department of Industry, Science and Resources (DISR). (Following the Federal election in 2001 mentioned earlier the Government portfolios were rearranged, and the CRC Program is now administered by the Department of Education, Science and Training. As the previous structure was in place for the bulk of this study, it will be referred to here.) The Program is overseen by the CRC Committee, a group of eminent scientists appointed by the DISR. This Committee is responsible for selecting those CRCs that are successful in the competitive bid process, and play a role in the continuing reviews over the life of each CRC. The Committee is supported in its role, as are the CRCs themselves, by the CRC Secretariat, a section within DISR.

Relations between the individual Centres and the CRC Program and Secretariat are supported and mediated by a Centre Visitor. The Centre Visitor is an eminent scientist in the field the CRC is working in. The primary role of the Visitor is to act as a mentor to the CRC, to assist in its development both technically and administratively, particularly with respect to meeting the objectives and following the processes of the CRC Program.

In addition, the CRC Association is a small lobbying body funded through levies on most CRCs. The CRC Association promotes the CRCs as a group, lobbying government for continued support and publicising the successes of the CRC Program. The Association also runs an annual conference that showcases the CRCs, and gives awards for excellent performance.

Organisational structure of CRCs

Legally, CRCs are most commonly unincorporated joint ventures among core partners. Core partners are contractually bound both to the Commonwealth and to each other. The Commonwealth Agreement is standard and covers main research activities, financial contributions, performance indicators and milestones (Australian Government Solicitor, 1998). The Centre Agreement is worked out by the participants, and covers issues such as conditions of employment, ownership of intellectual property, commercialisation of results, and disposal of assets (CRC Program, 2001c).

It is strongly recommended through these contracts that CRCs adopt a corporate-style of governance, with the appointment of a governing board to oversee operations. The Board "regulates all operations of the Centre including determining strategic development, reporting to the Commonwealth Government, approving projects, the annual budget and financial arrangements." (Coastal CRC, 2000, p. 5). The Board generally interacts with the Centre via the Chief Executive Officer (CEO) or Director. Generally, each core partner has a representative on the Board, and independent members may also be appointed.

The core partners of the CRC most typically include a mixture of research organisations and 'users', with the CRC Application Guidelines requiring strong interaction across a number of traditional boundaries:

> the development of effective collaborative arrangements is a key element in the success of a CRC proposal. The CRC should establish strong interactive linkages among individual researchers, between the participating organisations and between researchers and the users of research. This can be best achieved if the researchers from all the participating organisations in the CRC, including the user groups, are actively involved in a majority of the CRC's programs, and this is strongly encouraged (CRC Program, 1999, p. 8).

However, who those users might be remains deliberately vague, and the potential significance of differences in users remains largely unexamined:

A broad definition of research 'users' is intended. For example, a user might be an industry sector, a business enterprise, a rural industry research and development corporation, or a government department or agency responsible for areas such as the environment or resource management. (CRC Program, 1999, p. 9)

As such, there is considerable flexibility as to who users are and how they may become involved in CRCs.

Assessment and accountability There is less flexibility in the area of reporting. Indeed, the Federal Government describes rigorous reporting as a trade-off between freedom and public accountability:

CRCs are encouraged to adopt the operating procedures of a small company with a governing board, chief executive officer and line management structure.... In exchange for this strategic and management flexibility, CRCs are required to be accountable to the Commonwealth Government for the contribution of public funds. Consequently, independent reviews are held after year two and year five. (CRC Program, 2001, p. 3)

All CRCs have formal reporting and review obligations that are laid out in the Commonwealth Agreement contract. These reports are taken seriously, as failure to provide them constitutes a breach of contract with the Commonwealth, potential grounds for the Commonwealth to terminate the contract (Australian Government Solicitor, 1998, Section 15.1(e), 15.1(ba)). Hence the reporting and review system is an integral part of CRC activity.

Each CRC contract has a schedule attached that specifies the initial projects to be carried out and the milestones against which progress within these projects is to be measured. These milestones, as formal accountability structures, form the backbone of CRC research activity, as well as covering additional tasks such as publicity and communication, education, and business development. Each quarter, every project must provide a summary statement to the CEO, to be passed on to the governing board, which details expenditure (Australian Government Solicitor, 1998, Sections 13.1 and 13.2) and progress against the milestones specified in the contract. A sample of milestones is given in Table 3.1.

In addition, each Centre is also required to provide annual and final reports to the Board and the CRC Secretariat, for submission to the CRC Committee. These must include full details of the activities of the Centre, including "progress in the areas of research, education and training, collaboration and the use, involvement, commercialization and the application of research results generally", as well as "detailed information required for the evaluation of the Centre's performance in terms of the performance indicators", and "in terms of milestones." (Australian Government Solicitor, 1998, Section 14.1). This is further specified in a 16-page guide to the preparation of the annual report, which covers minimum requirements, from length and appearance to content and structure (CRC Program, 2001).

Table 4. A sample of CRC project milestones.

Milestones	Achievements	Milestones for 2000/01	Achievements 00/01
Database of the quantity and nature of forest products manufactured in Australia*	Existing databases reviewed Collaborative links established with sectors of the forest products industry	Database completed	Database will continue to be developed as gaps in current databases are identified and additional forest products industry data become available
Collection and analysis of [Western Australia] rangeland samples**	Woongan Hills and Moora samples collected, analysed for total organic C [carbon], particulate organic C and clay associated C	Further sites from WA identified in conjunction with [Conservation and Land Management]	Sites discussed but no further sites could be identified; three sites from [South Australia] sampled for future analysis

* Greenhouse Accounting CRC Project 2.3 ** Greenhouse Accounting CRC Project 3.2
(Source: CRC for Greenhouse Accounting, 2001b, pp. 58, 60 respectively)

At the same time as the annual report, researchers must also prepare a report that details any commercialisation or other exploitation of intellectual property. This must include reference to how the commercialisation or exploitation is consistent with the objectives of CRC program, and any benefits accruing to Australian industry, the Australian environment and the Australian economy generally. (Australian Government Solicitor, 1998, Section 14B). Further, CRC Chief Executive Officers are also required to complete and submit a management data questionnaire at the same time as the annual report is prepared. It requires largely quantitative information regarding the CRC's activities with respect to five categories of performance indicators: research; education; external communications, which includes publicising the Centre; commercialisation/technology transfer; and administration.

Major reviews are conducted at the end of each Centre's second and fifth year (CRC Program, 2001a, 2001b). They are also formalised in the Commonwealth Agreement contract (Australian Government Solicitor, 1998, Section 3.6). The reviews are carried out by independent reviewers in a process supported by the Centre Visitor. While the second year review theoretically has the 'teeth' to result in the Commonwealth withdrawing funds from the CRC, in practice this review is used to identify and fix any problems before they reach the need for such drastic measures. In contrast, the fifth year review is to assist in either developing a good rebid proposal for a second round of funding, or to ensure the steps are in place for wrapping up the CRC at the end of seven years. Fifth year reviews were not a part of this study.

The accountability structures within which the CRCs operate are significant with respect to this study, as it is through these procedures that the idea of 'integration' is formally assessed.

Integration in CRCs

As noted at the start of this chapter, the case studies I selected needed to be implementing some concept of integrated research within an environmental context. In my search for appropriate cases I needed to be able to justify my selections with respect to that criterion. Indeed, the best cases for my purposes would be those groups who were actively trying to implement the concept, not those who are paying lip service to some vague notion of integration to make a funding application sound more appealing. In this sense, self-ascription by researchers— simply saying that they were doing integrated research—was not sufficient as a basis for selection. Yet without a hard-and-fast definition of integrated environmental research against which I could judge claims of integration, how could that selection be made? The CRCs offered a solution to this problem. As their organisational context discussed above indicates, CRCs are formally required to work in partnership across organisational and institutional boundaries. While there are no guarantees as to the degree of interconnection, the competitive bid process and ongoing assessment were perhaps the best available indicators that CRCs were likely to be attempting to work together in the active sense.

There are three formal mechanisms by which the CRC Program encourages integration among the partners within a CRC: prior to formation, through application processes; at formation, through legal contracts; and following formation, through review processes.

Application Guidelines To be successful, CRC applications must meet the Application Guidelines. These highlight the importance of integration, stating that:

> the CRC should result in substantial integration of research activity that goes beyond the existing research efforts of the individual participants. (CRC Program, 1999, p. 6)

and further:

> Participants should form collaborative relationships within an integrated research program. Participants should not divide the research program into discrete projects that are carried out solely by individual participants, pursuing their own separate objectives. (CRC Program, 1999, p. 8)

These are perhaps the most readily enforceable criteria. As noted earlier, CRCs are awarded on the basis of a competitive bid process. Competition is usually fierce, as the seven-year funding contract is attractive both for its level of funding (the average CRC contract is somewhere in the order of AU$16 million of government funds over seven years, which is usually at least matched by the partners), and for the longevity of funding. With much research moving to a short-term contract basis, securing CRC support offers seven years of funding stability.

Consequently the Application Guidelines must be followed quite strictly to gain the Committee's approval. In 2001, ten years on from the start of the CRC program, a representative of the Committee noted that the quality of applications had steadily increased over that period (Brennan, 2001)—in other words, the competition was getting harder, and applicants were adhering more and more closely to the Application Guidelines.

Contractual arrangements Following the success of the bid, integration is reiterated more formally. As mentioned earlier, CRCs are formed by two contracts, one between the new Centre and the Commonwealth, and the other between the parties. In the former contract, there are two key objectives with respect to integration. The first is:

> To ensure that the Parties with their differing disciplines and background will, through their participation in the Centre, add value to each other so that the performance of the Centre will be greater than that of each Party performing independently. (CRC Program, 2001c, p. 6).

Secondly, the Centre must "promote the objectives of the Program." That is, must aim:

> [to strengthen] the links between research and its commercial and other applications, by active involvement of the users of research in the work... of the Centres;

> to stimulate a broader education and training experience... through initiatives such as involvement in major cooperative, user oriented research programs;

> to promote cooperation in research and through it a more efficient use of resources in the national research effort by building centres of research concentration and strengthening research networks. (Australian Government Solicitor, 1998, p. 2)

Whilst the terminology is necessarily vague in terms of action, the contract is a clear indication that the Centres need to take the collaborative, cooperative nature of the research seriously. This allows flexibility according to different circumstances, but also requires the Centres to effectively define their own criteria to justify their activities as 'integrative'.

Second Year Review These issues are followed up in the second year review, where the CRCs are assessed on:

> The degree to which key user groups, including industry, have been integrated into the CRC as core participants, and have made substantial commitments of resources [and] The degree to which the CRC has built links between the participating research groups and organisations, and integrated and enhanced their activities in research and education. (CRC Program, 2001b, p. 14)

The second year review is an assessment of how the requirements of the Application Guidelines and the contractual obligations are being implemented. It acts as a check against any temptation to give 'lip service' to the idea of integration.

In terms of this study, then, the CRCs had a clear mandate from their major funding body to apply the concept of integration—although it was not a formulaic approach. Different Centres were free to interpret and justify this 'integration' according to their own circumstances. Importantly, the formal requirements specified integration both 'within' the research, and between the research sectors and 'end users', and as such took a broader view of integrated research than interdisciplinarity. Further, the ongoing reporting and assessment criteria suggested that a 'one off' justification to get the initial CRC funding would be insufficient. Integration was an ongoing issue.

To this point, the context of the CRCs as a whole has been presented. This study focused on two CRCs, on the presumption that (given the context of CRCs in general) the activities of the environment sector CRCs would fall within the remit of 'integrated environmental research'. Two cases allowed for some diversity of context at the organisational scale, while maintaining the depth of interactions with each—given the resources available, more cases would have led to a more superficial involvement, inappropriate to the methodological aims of the study. Further, sharing the same organisational background set some boundaries around the extent of 'contextual difference' between the two cases—a degree of shared history between them provided some contextual similarity beyond their being concerned with environmental issues, broadly speaking. For the same reason, both cases selected were the same age: they were each funded in the 1998 funding round. This meant any comparison between the two organisations was based on a readily identifiable degree of similarity—comparing apples with apples, rather than apples with pears in the great fruit basket of integrated environmental research. Yet these apples were by no means of the same variety.

The two CRCs selected for this study were the CRC for Coastal Zone, Estuary and Waterway Management (Coastal CRC), and the CRC for Greenhouse Accounting (Greenhouse Accounting CRC). While more detail of the rationale for their selection will be given at the end of this chapter, in summary they were chosen for the diverse socio-political contexts in which they were situated.

THE CRC FOR COASTAL ZONE, ESTUARY AND WATERWAY MANAGEMENT

The Coastal CRC was based in Brisbane, Queensland, with central offices hosted by the Queensland Government's Department of Natural Resources and Mines. The Centre officially commenced operations in July 1999. It was difficult to gauge how many people actually worked in the CRC, as the boundaries of the organisation were permeable. Although the CRC directory included around 300 listings, most of the research staff had only part of their time allocated to the CRC as an in-kind contribution from their parent organisation. According to official figures, the full-time equivalent of staff in the Coastal CRC was 50 (CRC Program, 2002).

However, this figure was fluid, as different researchers were 'active' or 'inactive' according to the demands of their projects and non-CRC work arrangements, and different people were being brought in informally to fill small gaps in projects. In addition, several stakeholders committed significant amounts of time to CRC activities, but did not necessarily have an 'official' time allocation.

Mandate and rationale

The Coastal CRC was funded on the basis of it being a 'bridging' organisation, linking many stakeholders to bring about better coastal management in Australia. In their proposal, their aim was stated as:

> To bridge the gaps between science and the community, and between science and decision-making, policy and planning in the coastal zone. (Anon, 1998, p. 1)

As such, there was an emphasis on integrating across different institutions right from the beginning. This was in recognition that there are many different authorities responsible for management within the coastal zone, as well as high levels of community interest and commitment to improved coastal management. The coastal zone is the most densely populated part of Australia, and attracts many visitors for holidays, sport and recreation. Consequently the number and diversity of individuals, groups or organisations with a stake in coastal management are both high.

The Coastal CRC was funded at least partially on the basis that there was no major policy arena for Australia's coastal management, and therefore coasts tended to "fall between the stools", and fail to receive the coordinated support that was needed. In other words, the Coastal CRC in part aimed to generate public and policy awareness, to generate demand for the science that was available. To do this, the Coastal CRC worked both towards community awareness and grassroots action, and across all three tiers of Australian government policy: local, State, and Federal.

Core partners

Legally the Coastal CRC is an unincorporated joint venture between 10 core partners (counting CSIRO as one, although different CSIRO Divisions were involved. CSIRO was introduced in Chapter 2, and is a large research organisation structured into a number of major Divisions, e.g. Land and Water, Sustainable Ecosystems, Plant Industry, and so forth). There were also 12 supporting or non-core partner organisations, discussed in the next section. The core partners are listed in Table 5, along with their approximate funding contributions. Core partners are required to make a cash contribution alongside any in-kind support they may wish to provide.

As Table 5 shows, although there were 8 core partners in the Coastal CRC, like CSIRO, some of the larger organisations had several sub-sections involved. Adding the CRC grant from the Federal Government, of AU$14 720 000, to the total cash and in-kind contributions listed in Table 5, plus other small sums, brings the total budget estimate for the Coastal CRC to approximately AU$68 million over its seven-year lifespan (Coastal CRC, 2001).

Table 5. Core partners in the Coastal CRC with approximate cash and in-kind contributions over 7 years. (Source: Coastal CRC, 2001)

Core partner organisations	Total cash contribution AU$000 (%)	Total in-kind contribution AU$000 (%)	Location
University of Queensland	853 (13.5)	7167 (15.8)	Brisbane
• Dept of Botany			
• Dept of Chemical Engineering			
• Dept of Geographical Sciences			
Central Queensland University	700 (11.1)	4163 (9.2)	Rockhampton, Gladstone
• Centre for Land and Water Resource Management			
Griffith University	350 (5.5)	5481 (12.1)	Brisbane
• School of Environmental Engineering			
• School of Applied Science			
• Australian School of Environmental Studies			
• School of Environmental Planning			
James Cook University	175 (2.8)	2102 (4.6)	Townsville
CSIRO	(1575 (25))	(10311 (22.3))	
• Division of Marine Research	700 (11.1)		Hobart
• Division of Land and Water	700 (11.1)	4553 (10.1)	Canberra
• Division of Maths and Information Technology	175 (2.8)	3487 (7.7)	Brisbane
		1082 (2.4)	
• Division of Energy Technology		1189 (2.6)	Sydney
Queensland Government	(1720 (27.2))	(11465 (25.5))	
• Department of Natural Resources and Mines	700 (11.1)	5436 (12.0)	Brisbane
• Department of Primary Industries	245 (3.9)		Cairns
		1815 (4.0)	
• Queensland Environment Protection Agency	775 (12.3)	4214 (9.3)	Brisbane, Rockhampton
Brisbane City Council	350 (5.5)	2116 (4.7)	Brisbane
Geoscience Australia	600 (9.5)	2488 (5.5)	Canberra
Totals	6 323 (100)	45 293(100)	

Non-core partners and the National Stakeholder Advisory Committee

Non-core partners include several small to medium-sized enterprises and several industry bodies. Unlike their core counterparts, they are not required to make a cash contribution to the CRC, but often provide in-kind support. Non-core partners maintained an active interest in the CRC, some as potential technology advisers and suppliers (for example, Australian Interactive Multimedia Association, Netstorm), others as stakeholders wishing to both inform and be informed of research progress, such as Douglas Shire Council (a local government authority), Gladstone Port Authority (a government-owned corporation), and Southern Pacific Petroleum (a private company).

Many of these stakeholders were represented on the Coastal CRC's National Stakeholder Advisory Committee (NSAC). This group was formally established to contribute to the CRC's operations, from strategic planning through to project selection and development (Coastal CRC, 2001, p. 5).

Organisational structure

The organisational structure of the Coastal CRC is illustrated in Figure 8.

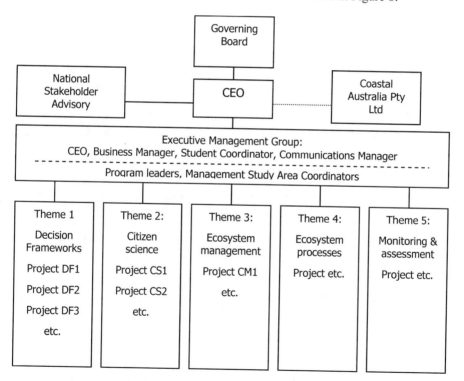

Figure 8. Organisational structure of the Coastal CRC.

The main features of the organisational structure of the CRC include the Board, who held legal responsibility for oversight of the CRC; the Chief Executive Officer (CEO) who managed the day-to-day affairs of the Centre, and was directly responsible to the Board; the National Stakeholder Advisory Committee (discussed above) and the thematic research structure. The Coastal CRC chose to describe their research clusters as 'themes' rather than the more conventional descriptor, 'programs'. Coastal Australia Pty Ltd was the commercial and business arm of the Centre. It was not yet fully operational when this study commenced, as was therefore not a key part of the research. The CEO, Theme leaders and other key personnel formed the Executive Management Group, who met regularly via teleconference to report on and plan the Centre's operations.

The Board   In the Coastal CRC the Board consisted of four independent and eight partner members. As mentioned, each partner to the CRC may nominate a Board member. Representatives tend not to be actively involved in the research programs, as they are usually more senior than the research staff, to ensure that the Board member can make decisions at the Board table, rather than having to gain approval through their own organisation's hierarchy. In the Coastal CRC, due to the large number of partner organisations, to keep the Board membership to a workable level there were two representatives from the four Universities who were rotated on an annual basis.

The Chair of the Board was a high-profile figure in science policy both nationally and internationally, and was an independent Board member, not formally affiliated with any of the partners. The Deputy Chair was also independent, with a background in banking rather than research. While their interaction with the CRC was primarily through the CEO, the CEO also provided other opportunities for staff and students of the CRC, as well as other parties such as the National Stakeholder Advisory Council, to meet directly with the Board.

Chief Executive Officer (CEO)   The CEO of the Coastal CRC at the time of this study was Dr Roger Shaw. Prior to holding this position with the CRC, Dr Shaw had led the Queensland Government's Department of Natural Resources Strategic Science Initiatives program. As CEO he was ultimately responsible for the operational management of the CRC, and was directly responsible to the Board. There was no deputy CEO position.

Executive Management Group   The CEO also led the Executive Management Group, the composition of which is indicated in Figure 8. The Executive Management Group included key support staff, such as the Business Manager, Communications Manager, and Education Coordinator. These people were employed by the CRC (100% of their time was devoted to the CRC). They played significant roles in the CRC as they not only cut across the research groupings (discussed below), but also were key managers of relationships between the CRC and the 'outside world' as well. This was particularly the case for the Communications Manager, who was responsible for internal communications and external public relations as well as organising activities such as media training for researchers. Likewise the Business Manager was central to developing and maintaining good business relationships with external non-core partners and other providers, and the Education Coordinator with university representatives who may or may not be part of the CRC.

Research structure

The research was carried out in projects (many of which were broken down into tasks), which were placed in one of five research themes. Each theme had a leader who reported to the CEO. An early decision by the CEO required that all projects have representatives from at least two partner organisations.

The themes ("programs" in the Greenhouse Accounting CRC) formed the main research structure of the CRC. They were not discipline-based, but rather organised according to what might best be described as their scale of management relevance. The five themes were: Monitoring and Assessment (managing individual species); Ecosystem Processes (managing interrelations between species); Planning and Management (managing biophysical ecosystems); Citizen Science and Education (managing people in ecosystems); and Decision Frameworks (managing people and ecosystems together). The five themes were habitually presented as illustrated in Figure 9. In this representation of the research structure, the theme at the base of the triangle feeds into the next level up, and so on, in increasing levels of complexity until the final theme, decision frameworks, incorporates information from all of them. For a more extensive discussion of the research themes, see van Kerkhoff (2005b). While the 'integrative' aspects of this structure will be discussed in more detail in Chapters 4 through 7, it was promoted by the CEO both within and outside the CRC as the centrepiece of their research. It appeared, for example, on the reverse side of all CRC business cards and on printed CRC 'stubby-holders'

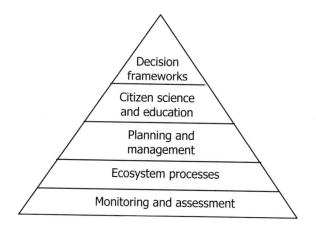

Figure 9. The "Shavian Triangle". (Source: Coastal CRC, 2000)

(neoprene drink coolers). The "Shavian Triangle" was an affectionate title, as the CEO Roger Shaw was its devisor and keen promoter, especially during the early phase of the CRC.

Themes were coordinated by Theme Leaders. The Theme Leaders in the Coastal CRC were required to have at least 50% of their time devoted to the CRC. Their role was to provide scientific, logistical and operational support and coordination for the researchers in their Themes, and represent them in the Executive Management Group meetings. They were also expected to coordinate activities with other Theme Leaders and to develop better synergies between and across themes.

The thematic structure was complemented by a second structure: Management Study Areas (MSAs). These where physical locations where the CRC concentrated its research effort. The four MSAs were South East Queensland (rivers and estuaries of Brisbane and Moreton Bay), Fitzroy (Central Queensland, based in Rockhampton), and Gladstone (an industrial port on the Calliope River, also Central Queensland). These represented different major environmental use patterns: metro-politan residential and recreational; agricultural; and industrial, respectively. There were also a small number of nationwide research projects that didn't really fit the idea of an MSA representing a single locality, but were nonetheless granted "national" MSA status.

The locations of the MSAs highlighted that the majority of the CRC's work took place in the state of Queensland, as illustrated in Figure 10. This rather unbalanced

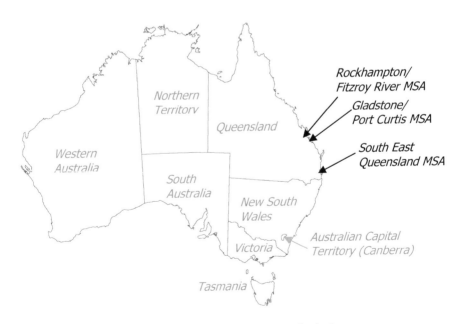

Figure 10. Location of Management Study Areas.

approach reflected that the CRC's connections and networks were strongest in their home state, despite drawing on researchers from other parts of the country including Canberra and Tasmania. However the concentration of work in Queensland was widely recognised as problematic in the CRC, as CRCs in general were supposed to serve national interests, rather than State ones. Accordingly, over the course of this study the CRC attempted to develop links with areas outside Queensland, none of which had been successful by the time the study concluded in 2002.

The MSAs were also an organisational unit within the CRC, each with a coordinator whose role was to oversee research projects in their area (including facilitating access to resources, etc.) and to link the CRC with local (or national) stakeholders. MSA coordinators also held a position on the management team, with the same managerial responsibility as Theme Leaders.

The overlay of physical locations over themed groupings was described by one participant as a matrix arrangement, where research activities took place simultaneously in two dimensions, as represented in Figure 11. However, while all projects were explicitly located within a theme, they were not necessarily likewise formally associated with a MSA, and not all researchers identified with both structures.

Theme/MSA	SE Queensland	Fitzroy	Gladstone	National
Decision frameworks				
Citizen science				
Planning and management		Project x		
Ecosystem processes				
Monitoring and assessment				

Figure 11. A matrix of Themes and Management Study Areas.

Socio-political context

There were many factors associated with the Coastal CRC's socio-political relevance. One was that there are many government and other statutory bodies responsible for coastal management in Australia (Resource Assessment Commission, 1993). Another factor was that the pressures on the coastal zone are large and increasing. As the Resource Assessment Commission wrote in their Coastal Zone Inquiry in 1993 (and was often cited within the CRC):

The coastal zone has a special place in the lives of Australians. Most Australians want to live there and if they can't they want to take their holidays there. It contains diverse ecosystems and a high proportion of

>Australia's industrial activity occurs in the zone. It is a priceless natural
>resource. (Resource Assessment Commission, 1993, Section 2.01)

This quotation points to a further issue—coasts and, in particular, beaches are an integral and important part of Australian culture and identity. As such there was pressure and demand for solutions to coastal management, as well as newly legislated requirements for environmental protection monitoring and auditing of the current health of estuarine systems through a major National Land and Water Resources Audit.

In political terms, coastal management was a highly dispersed issue. The number of political agencies with responsibility for coastal management was high, and included local government councils, who are responsible for water supply and sewerage as well as coastal development; State government agencies, including environment protection authorities and natural resource management agencies with a wide range of legislative responsibilities in the coastal zone; and Federal Government departments, including Environment Australia, and Agriculture, Fisheries and Forestry Australia, as well as Federal statutory authorities such as the Great Barrier Reef Marine Park Authority. This complex situation was coupled with increasing pressure on industry to reduce their impact on coastal environments (from any or all of these government sources, as well as public pressure), and increased community awareness and activity in water quality management (Waterwatch, Sea-grass Watch, etc.). In short, the number of stakeholders was high and their backgrounds and interests were diverse.

The Coastal CRC, through their aim to 'bridge the gaps' between these groups, could not simply choose to work with a handful of these organisations and ignore the rest. They needed to be able to somehow incorporate this diversity into their operations. Consequently, much of the Coastal CRC's credibility centred on their accessibility across this range of players. The CRC's ability as a group of scientists was not in question—it was their ability to generate meaningful research and communicate its relevance that was the major political driver for their development.

This socio-political context is significant in terms of this study, as the Coastal CRC's background serves as the basis for the *idea* of integrated research as it was applied in this Centre. In other words, the questions of integration of what? With whom? For what purpose? and so forth, emerge from the dynamic between the core of the CRC and this periphery. As noted in opening this chapter, this is not a 'context' in the sense of something separate from 'the research', but a vital—a living—part of the day-to-day practice that made up what it was to be doing integrated environmental research.

THE CRC FOR GREENHOUSE ACCOUNTING

The CRC for Greenhouse Accounting was based in Canberra, at the Australian National University, 'hosted' by the Research School of Biological Sciences (RSBS). The Centre was officially opened in December 1999.

The scale of research in this CRC was difficult to gauge according to staff numbers, with the issues faced in the Coastal CRC largely applicable in the Greenhouse Accounting CRC. Most researchers split their time between CRC work and other work, and the work arrangements were fairly fluid. Overall, the research team of the Greenhouse Accounting CRC was slightly larger than the Coastal CRC—in 2002 their full-time equivalent staff numbers was reported as 57 (CRC Program, 2002).

Mandate and rationale

The main objectives of the CRC for Greenhouse Accounting were to:

attempt to understand the Australian terrestrial carbon cycle and how it responds to global climate change. To do this, the Centre will research and develop innovative, cost-effective methods for land management and accurately measuring and forecasting change in land-based carbon stocks.

The Centre will also help devise and promote modern tools for managing land-based carbon so as to help achieve national greenhouse gas reduction objectives. (CRC for Greenhouse Accounting, 2001a, inside cover)

Like the Coastal CRC, the Greenhouse Accounting CRC was also largely about coordinating and bringing together a wide range of skills, expertise and data that were relevant to carbon accounting, but dispersed over several agencies. The CRC's agenda was closely tied to that of the Australian Federal Government, as climate change negotiations were in motion, and the science of greenhouse accounting— measuring carbon in forests and other land uses, and how it changes—was recognised as crucial to Australia's response to greenhouse issues. This placed considerable demand on the relevant science communities to feed into policy processes.

Core partners

Legally, the Greenhouse Accounting CRC is an unincorporated joint venture made up of 8 core partners. These are listed in Table 6, along with their approximate cash and in-kind contributions.

Adding the Federal Government's CRC grant of AU$15 360 000, plus other small sums brings the Greenhouse Accounting CRC's total estimated budget over the seven years to around AU$69 million. This does not include income from future contract-based work. Hence the budgets of the two CRCs are very similar in scale, but quite different in composition. Although they have the same number of core partners when the Queensland Government and CSIRO are counted each as single units rather than by Department or Division, when those separations are taken into account, the Coastal CRC partners outnumber the Greenhouse Accounting CRC's partners by 13 to 10 (The same comparison across university departments cannot be made because these figures are not broken down in the Greenhouse Accounting CRC's reporting.). Another key difference is that the Greenhouse

Accounting CRC cash contributions are dominated by a single organisation, the Australian Greenhouse Office, which is not an active research partner.

Non-core partners The Centre also had six supporting partners: Vice Saltbush Company of Australia; Alcoa of Australia; Stanwell Corporation; Shell Company of Australia, Chemistry Centre WA, and the University of Melbourne. The latter two provided in-kind support for the CRC; the four private companies committed to cash contributions of between AU$60 000 and AU$150 000 over the course of the seven years.

Table 6. Core partners in the Greenhouse Accounting CRC with approximate cash and in-kind contributions over 7 years. (Source: Greenhouse Accounting CRC, 2001)

Core partner organisations	Total cash contribution AU$'000 (%)	Total in-kind contribution AU$'000 (%)	Location
Australian Greenhouse Office	1188 (43.0)	0	Canberra
Bureau of Rural Sciences	175 (6.3)	5601 (11.1)	Canberra
Australian National University	700 (25.3)	22067(43.9)	Canberra
CSIRO			
• Division of Plant Industry	0	8526 (17.0)	Canberra
	0		Canberra
• Division of Land and Water			
Queensland Government		5975 (11.9)	
• Department of Natural Resources and Mines	0	3154 (6.3)	Brisbane
• Department of Primary Industries	0	2821 (5.6)	Brisbane, Rockhampton
Western Australia Chemistry Centre and University of Melbourne		799 (1.6)	
Department of Conservation and Land Management Western Australia (CALM)	0	2427 (4.8)	Perth
State Forests New South Wales (SFNSW)	700 (25.3)	4900 (9.7)	Sydney
Totals	2763 (100)	50 295 (100)	

Organisational structure

The organisational structure of the CRC was similar to that of the Coastal CRC, as illustrated in Figure 12. This structure changed in 2001 following an internal strategic review, which will be discussed further later.

The Board The Greenhouse Accounting CRC Board was comprised of 12 members, 4 of whom were independent, including the Chair.
There was no Deputy Chair position for the Board.

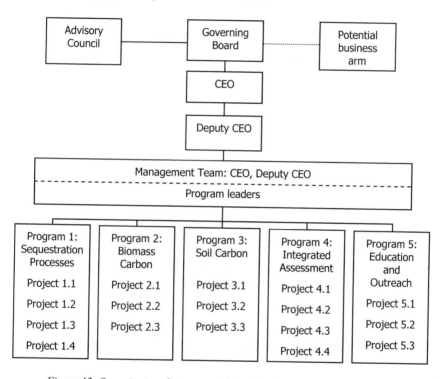

Figure 12. Organisational structure of the Greenhouse Accounting CRC.

The Advisory Council The Advisory Council was comprised of a range of interested stakeholders. It was intended to provide "a forum for user and advisory groups as well as government and non-government agencies to provide input into the Governing Board's decision- and policy-making" (CRC for Greenhouse Accounting, 2000 p. 54). The first Advisory Council meeting was held in July 2001, and so they had only a limited role over the course of this study.

Although similar to the Coastal CRC's National Stakeholder Advisory Committee, the Advisory Council reported directly to the governing board, rather than through the CEO.

The CEO The CEO of the CRC was Professor Ian Noble from the Research School of Biological Sciences at the Australian National University. Professor Noble was a member of the International Geosphere Biosphere Program (IGBP) and chair of the Global Change and Terrestrial Ecosystems committee within that program. As such he had played a prominent role in the Intergovernmental Panel on Climate Change (IPCC) Scientific Committee, and was a key adviser to the Australian Government and international organisations on climate change issues. He continued to play a major role in the international arena throughout his time with the CRC. In 2002 he resigned as CEO, and at the time this study concluded there had been no permanent replacement made, with the Deputy CEO acting in the CEO position.

Executive Management Team The day-to-day operations of the Centre were jointly managed through the Executive Management Team, comprised of the CEO, Deputy CEO, Program Leaders, and Business, Operations, Communications and Education Managers.

The research structure

The original research structure of the CRC was based on four programs, roughly equivalent to the Coastal CRC's research Themes. Each program was comprised of a number of projects. There was also a fifth supporting program, Education and Outreach, although it was responsible for postgraduate students, other educational activities, and public communications rather than direct research. This structure changed part way through this study (which will be discussed in detail in Chapter 6), and the original structure is reported here. The four research programs were Sequestration Processes; Biomass Carbon; Soil Carbon; and Integrated Assessment.

The first three programs were comprised of basic and applied research projects, concerned with different aspects of measuring carbon in plants and soils. Together they were designed to feed information into the fourth program, Integrated Assessment, particularly Project 4.1, Integrative modelling. As the title implies, this project was tasked with bringing the disparate research outputs together into a single, integrated model. This project, along with 4.2, was designed specifically to "assist the future development of the National Greenhouse Gas Inventory (NGGI)..." (Greenhouse Accounting CRC, 2000, pp. 24–25). The Inventory was being compiled by the Australian Greenhouse Office (AGO) to establish Australia's baseline in carbon accounting—essentially, to work out how much carbon is in the country. Similarly, the National Carbon Accounting System was a methodology being compiled by the AGO. Although these were being overseen directly by the AGO, rather than by the Centre, the CRC's research was seen as providing a basis for their longer-term development. These relationships are illustrated in Figure 13.

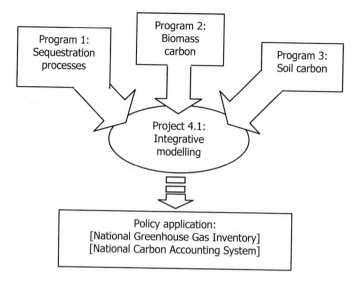

Figure 13. Research flows in the Greenhouse Accounting CRC.

All projects were subject to initial milestones as laid out in the Business Plan attached as a Schedule to the CRC Proposal. These were also revised over the course of the strategic review.

The support staff Other major staff positions at the Greenhouse Accounting CRC included the Business, Operations, Communications and Education Managers. As in the Coastal CRC these positions were significant in the CRC because they cut across the research programs. There were postgraduate students in most programs, every project had an obligation to see their research 'communicated' in some way, often by working with the Communications Manager, and so on. These staff were also responsible for developing good 'external' relationships, for example the Education Manager developed short courses on carbon accounting for professional development to external parties, and so on.

Socio-political context

The Greenhouse Accounting CRC differed substantially from the Coastal CRC in its political and social background. At the time of its formation, several major political forces were developing in both the national and international political arenas. At the international level, the Kyoto Protocol negotiations were in full swing.

The Greenhouse Accounting CRC was also quite strongly 'science specific' as it was concerned with a particular aspect of climate change—indeed, two particular Articles of the Kyoto Protocol. Articles 3.3 and 3.4, and later article 3.7 (the so-called 'Australia Clause') of the protocol were concerned with the role of carbon sinks (mostly trees and forests) in accounting for greenhouse emissions and reductions. These articles are listed in Appendix 2. At the time of funding (mid-1999), several CRC researchers were heavily involved in contributing to the Intergovernmental Panel on Climate Change (IPCC) *Special report on land use, land use change and forestry 2000* (Watson et al., 2000) as well as the Third Assessment Report, *Climate change 2001: the scientific basis* (Houghton et al., 2001), a major international document detailing the state of scientific knowledge about carbon sinks and their role in climate change abatement. This was a highly political issue in Australia, as the Federal Government's position at the Kyoto negotiations argued strongly that sinks should be included in calculations of how far a country would be required to reduce its carbon emissions. This was in large part based upon the belief that Australia, with a large capacity for plantation afforestation, would benefit from the inclusion of carbon sinks if large-scale forest regrowth counted as carbon credits. This would reduce the impact of emission targets on other economic sectors and place Australia in a strong position in terms of international carbon trading. This also then emerged as a key topic for State governments, as the States are largely responsible for land management, including forestry activities.

The Australian Greenhouse Office, mentioned earlier as a key funding partner and research user, was part of Australia's response to the international negotiations. The Federal Government created the Australian Greenhouse Office as a new agency in 1998, to act as a central point for all greenhouse-related issues. The Australian Greenhouse Office spanned several existing government departments and agencies who each were responsible for different aspects of climate change, including Environment Australia; Agriculture, Fisheries and Forestry Australia; Transport and Regional Services; and the Department of Foreign Affairs and Trade. As mentioned earlier, part of the Australian Greenhouse Office's mandate was to develop the National Greenhouse Gas Inventory, mentioned above, and a system for accounting for the nation's greenhouse gas emissions and sinks—the National Carbon Accounting System (NCAS). Several key players involved in the creation of the CRC were also heavily involved in the NGGI process, while others were appointed to the high-level Steering Committee for the NCAS. In contrast to these tasks that specifically drew on existing techniques and data, the AGO regarded the CRC as a medium- to long-term investment in the conceptual and technical development in the area of land management carbon accounting.

However, the tools available for measuring carbon in plants were limited. Hence the Greenhouse Accounting CRC was focused primarily on issues concerned with understanding carbon sequestration in plants, particularly forests. So, the issue was relevant at state, national and international political scales, but to relatively limited and clearly defined groups: State Government forestry agencies and natural resource management/land management departments, and the AGO in particular. This high-level, quite concentrated focus of interest was a stark contrast to the local, dispersed groups with an interest in coastal management.

This may be related to the science-specificity of research within the Greenhouse Accounting CRC. Although a broad and complex field in the sense that terrestrial carbon accounting drew upon many disciplines and organisations, the role of the Greenhouse Accounting CRC was conceptually quite targeted. This was again in contrast to the Coastal CRC, who did not restrict the conceptual range of their research activities to particular aspects of coastal management.

THE RED DELICIOUS AND GRANNY SMITHS OF INTEGRATED ENVIRONMENTAL RESEARCH

Consequently, this study was comparing apples with apples, rather than apples with oranges—but these apples were of very different varieties. The contrast between the two socio-political contexts created the main comparative basis across the two CRCs. Organisationally, both Centres were quite similar: with relatively minor variations they operated along organisational structures that were much the same (an important difference was the Management Study Areas in the Coastal CRC, which will be discussed further in later chapters). Yet within the larger similarity of both CRCs being concerned with environmental issues and environmental management, the socio-political context of each was very different. As noted above, the Coastal CRC was immersed in a highly dispersed political and practical milieu, with many tiers of government, industry of various scales, and community groups with varying capability and resources. The Coastal CRC's mandate to 'bridge the gaps' between science and policy led them to engage with all these players. The issues of coastal management also tended to be location-specific, in the sense that it was concerned with the coastal fringe, and most commonly with particular parts of the coastal fringe (national survey projects notwithstanding) particularly via the MSAs. However, it was not, so to speak, 'science-specific', open to a broad spectrum of biophysical as well as social issues that related to coastal zone management, from detailed species-specific studies (counting marine pests) to social-psychological studies (community information-seeking behaviour) and high-level computer-based decision support systems.

The Greenhouse Accounting CRC, in contrast, was concerned with one specific aspect of the global issue of climate change: measuring carbon in managed landscapes. Hence their stakeholders (acknowledging that the entire global population is a stakeholder in climate change in the broad sense) were limited to those with a stake in a Australia's ability to count carbon—primarily Federal Government agencies, and some State agencies. In addition, the Kyoto Protocol was in a state of flux over most of the course of this study, placing scientists working in this area under considerable demands from policy-makers, as well as under considerable scrutiny by sceptics.

The character and diversity of the Centres' stakeholders combined with the different scopes of their scientific agendas, placed the Coastal CRC and the Greenhouse Accounting CRC towards the opposite ends of a socio-political contextual spectrum. These differences, along with the similarities of age and environmental focus, made these an unusual and exceptional pair of cases for exploring the articulation and conduct of integrated environmental research.

TALK OF INTEGRATION

In this chapter I will focus on the ways people made sense of integrated research as an abstract concept. This is the first step in a process of progressively unpacking the concept of integrated environmental research and its implications for what people say about it, how they do it, what can be learned from it, and what this all means for research practice. As Chapter 3 showed, CRCs needed to be able to justify their operations in terms of integration—this requires participants to construct an understanding of what integrated research *is*, in the abstract sense that could be included in reports and demonstrated to review panels. What dimensions of practice did these constructs tend to highlight?

FROM PRACTICE TO PRESENTATION

The concepts of integrated research presented in this chapter were drawn from the research conversations by close examination of the ways in which the participants used the actual word 'integration' (or its variations—integrated, integrating, and so forth). More specifically, the use of the term was examined *in relation to* the concepts and categories that were used alongside 'integration'. These categories indicated how integration was nested within larger conceptual schemas. These schemas formed the basis of the context within which integration—and the relationships it encompassed—gained meaning at particular points in the conversation, and became a source of understanding.

These understandings are important, as they constitute the participants' *abstract understandings* of integration, which influence the ways in which they learn from their experiences. In other words, the models presented here are not simply different views of 'what integration is', but represent the conceptual frameworks that shape how the participants *learn from* their practices of doing integrated research.

Consequently, in this chapter I will present fragments of conversations that are grouped together according to the conceptual schemas within which the use of the term 'integration' was embedded. Indeed, the way people talked about integration as an abstract concept illustrated several different, but related, constructs of what integration is, how it should be done, what it might achieve. They were often expressed slightly differently, but could be grouped into six general models. In this section I will introduce the models in turn, and then discuss some of the consequences that flowed from these categories.

They are presented roughly in order of increasing complexity, with the exception of the last model, as it is related to previous ones.

Model 1. Container

The container model was the least structured model, relying primarily on the assumption that if you place people with disparate information and knowledge together, they will interact to produce a more integrated outcome than if left alone. This model suggested that new knowledge is generated at least in part by serendipity and collegiality, and that the role of integrated research organisations is to create the container, to facilitate or even force that interaction.

In this model there were two main variables: what the container was, and what 'bits' of information were placed in that container as illustrated in Figure 14.

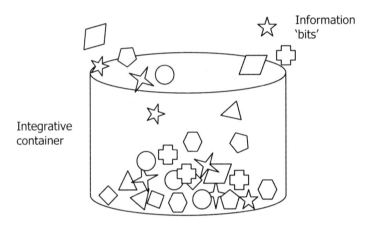

Figure 14. The container model of integration.

In the container model, information sharing tended to be regarded as a flexible process depending on good will and serendipity rather than slotting into a formal structure. While some structure (the 'container') helps to overcome the tendency for people to prefer to work in isolation, or in their comfortable home groups, within that fairly general arena people were free to move in directions that appealed to them.

The integrative container was sometimes perceived to be the CRC itself, with the 'bits' being different scientific processes that were usually kept apart. As one participant described it (participant quotations are italicised):

the fact that both [modelling and reductionist] scientific processes are embedded in the same CRC at least gives the best possible opportunity for collaboration and influence. That's certainly better than what occurred before which was where the two sides probably never communicated, and probably judged each other to either be irrelevant or too pragmatic to be useful.

Other types of 'container' were also possible. Physical spaces served as useful containers, as in the Coastal CRC's Management Study Areas. Most of the CRC's projects were carried out in these regions, on the basis that locating research in specific areas would allow greater collaboration with stakeholders. The Coastal CRC's annual report states:

> As a primary method to bridge the gaps between science and the community, policy, planners and decision-makers, the Coastal CRC adopted the concept of quality science with direct application. This was achieved by having projects address specific issues in management study areas that were identified in consultation with stakeholders, with the science relevant to those issues. (Coastal CRC, 2001, p. 23)

In other words, the container (in the form of a local region) restricted the number of stakeholders (bits) with an interest in the research processes, but maintained the CRC's ability to incorporate more types of bits, from community groups and local and state government to industry. While the Greenhouse Accounting CRC did not have the equivalent of Management Study Areas, it did seriously consider introducing core sites to serve a similar purpose, albeit among the research groups.

In terms of relationships, the containers in this model can be seen as opportunities to re-categorise people, or, at least, the groupings they represent, and use that re-categorisation to create new commonality between them. The quotes indicate that the importance of the *people* in the container lies in their being representatives of different relevant expertise or capacities: community views, scientific processes, disciplinary expertise, decision-making capacity, and so forth. The CRCs themselves were a new way of classifying one researcher's relationship to another. Whereas before one may have been a researcher in a small division of a State agency, and another a university professor on the other side of the country, now they were all CRC members. This legitimised contact:

> *I think what [the CRC] offers is unprecedented cooperation between partner agencies across state boundaries, across jurisdictional boundaries, across tiers, federal, state, local, across Australia, Perth to Rockhampton— unprecedented... you just have to pick up a phone and you have the weight of resources behind that, working for you, really.*

With contact legitimatised, people could work out their shared interests and develop work practices (such as project teams, steering committees) that capitalised on their new relationship. This is perhaps a rather romanticised view of collaboration and integration, where people somehow find each other across a crowded container, realise how much they have in common and develop new, exciting, productive relationships. One researcher drew this analogy quite clearly, with reference to his emerging relationship with his project team:

> *It's no longer like a sort of rather diffident courtship.... It's building understanding and trust, you can interact with people, you get to know them, but it's almost in a social sense rather than in a work related sense, I think anyway.*

Of course, as shown in Chapter 3, the CRCs were not an interactive free-for-all. They did have particular structures to bring some people together, which also kept others apart. In some cases, dissatisfaction with the CRCs was related to their tendency to 'force' people together who may not want to work together, while others could not work with those they would like to. Comparisons with past serendipitous ('romantic') enjoyable collaborations reminiscent of the container model were often unfavourable towards the CRCs. As one researcher expressed it "Life's too short to work with jerks". In contrast, however, other researchers felt that the CRC had presented them with a positive opportunity to meet and work with others whom they 'never even knew existed', or had 'known for years' but had never had the chance to work with. Consequently, the container model was occasionally used as a point of comparison with other collaborative relationships.

A consequence of viewing integration in this way was that common courses of action were basically variations of either the container, or the bits within it. It was versatile, in that it allowed for common containers to be created around a wide range of people, across several boundaries—scientist or non-scientist, modeller or empirical researcher, and so forth.

The other course of action within this model was to vary who was in and who was out. For example, one member of the Greenhouse Accounting CRC voiced concerns that the current range of people in their CRC container was insufficient to meet policy-driven goals:

> I think it's very natural to respond to the loudest voices that are closest to you, and I'm not belittling the importance of those voices but I think it's wise to be aware that there is a much wider constituency out there who have a vital interest in [greenhouse accounting] and if any of the… policy agendas are to be realised, [the CRC will] need to get out to this community as well.

A third course of action was to facilitate the bits in the container coming together to form productive relationships. The lack of structure in the container model meant that while the presence of a container such as the CRC might encourage people to work together, it did not insist that they integrate their work. Nor did it necessarily support those who needed additional resources to do so. To remedy this, in its second year the Greenhouse Accounting CRC introduced a 'collaboration fund', which aimed to facilitate integration within the CRC container, by providing resources to those who proposed collaborative research with others in the CRC, but outside their immediate groups. In some respects, this was an effort to reinstate the CRC *as a whole* as an effective container, over and above the other structures (such as programs) that were in place.

The container model highlighted an ongoing balancing act for the CRCs between the needs for structure and flexibility. While implementing structures such as programs, governing boards, advisory committees and so on were necessary to create clear pathways for integration, these pathways also served to close off other paths that people may have wished to explore.

To summarise the container model: containers are new categories that highlight or create commonality between people. This commonality facilitates spontaneous connections and collaborations leading to integrated outcomes. In many respects it is a model of integration that still allows for a classic ideal of collaboration: a serendipitous meeting that leads to mutually satisfying joint work. While the people in the containers are usually included for their representation of a desired group, in this model integration is primarily a human process, sometimes tinged with a somewhat romantic air. The main courses of action suggested by the container model are the creation of new containers, and expanding or contracting the number of people within them according to which interests or expertise were missing.

Model 2. Purchaser–provider

The 'purchaser–provider' way of talking about integration reflected, as the title suggests, a 'research as business' orientation. Research is a service to be provided on the basis of a financial, contractual relationship. This model was perceived to be integrative across science–non-science boundaries in the sense that the research was responding directly to an information need defined in non-science sectors. It could also be integrative across disciplines within science, as the work was usually highly problem-focused, and hence often demanded a multi-disciplinary approach.

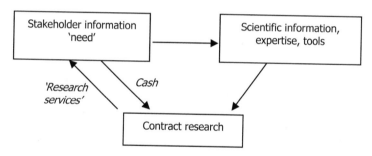

Figure 15. The purchaser–provider model of integration

The language of the purchaser–provider model was broadly drawn from the language of business: research groups were 'service providers', research outcomes were 'deliverables', and 'products', which were information. As such it was based on an assumption of transferability: that information generated in a scientific context would be meaningful when transferred outside that context.

The pressure for researchers to adopt this model of integration was, in some cases, intense. Not only were consulting monies brought into the CRC part of the CRC performance indicators, stakeholders who saw advantages in drawing on the combined expertise present in the CRCs through commissioned research could also push this model strongly. As one stakeholder described it:

We've been feeding proposals to the CRC... we are almost preferential, we're saying 'Look, this is something that's well-suited to the CRC, put together a bid, this is what we want and we are prepared to look at the CRC on a preferred provider basis'.

The Australian Greenhouse Office, for example, co-funded the appointment of a Business Manager position to 'kick-start' the Greenhouse Accounting CRC's capacity to engage in contract research.

The presence of contracts embedded the relationships between purchasers and providers in the legal-administrative system. Obligations were enforceable through formal legal structures, which in turn required accountability measures manifested in administrative processes such as budgeting and reporting. Often the purchaser did not get involved in the research, but set out the terms of the contract and then waited for the research service to be delivered. It was the purchaser's responsibility to determine what information they needed, and who could best provide that; and it was the provider's role to meet the specifications of the contract. As one purchaser noted, while some people accepted this categorisation, others did not:

> some people quite quickly understood that model and engaged and saw themselves fitting within that purchaser–provider framework. Others saw themselves as, I guess, not wanting to work under the strict delineation of 'you will provide to a specified product', prescription of activity. They tended to choose not to engage, and there's a mix of that within the CRC.

While the purchaser–provider model was a highly formalised structure, with the terms of the relationship literally laid out in black and white, seemingly immutable, there was still scope for uncertainty. For example, the Coastal CRC, while still engaging in contract-based research, preferred to view these relationships as joint partnerships, which ran contrary to a strong historical precedent: the contracting parties often had clear expectations on the basis of contracts they had held before, with other groups. One research manager found this frustrating:

> *that's the way they do business, they know what it means, they know how to manage it, they just expect you to do their bidding.*

This extends from the research into the administration of the research as well. The difficulties of grappling with recalcitrant data, or getting the right research team together, or adequate administrative procedures, are not part of the purchaser's purchase: why would you buy other people's problems? As one purchaser commented, it is their (the providers') responsibility to have the right staff in place who can do the job: "that's not our problem."

Issues of intellectual property were intertwined with the purchaser–provider model, with the potential for legal restrictions to be placed on information, preventing it being shared more widely. In the experience of most of the participants, intellectual property was rarely a serious threat to collaboration and integration, but it could hamper collaborative efforts:

> *I haven't seen any evidence of people involved not informally sharing information, like over a discussion, but certainly I've seen evidence of people saying 'well, I've done that piece of work but I can't give it to you because it's owned by a third party'....*

Similarly, where purchase of, or negotiation of access to, data generated and held by particular partners was involved, sharing could be difficult. This is noteworthy, as it indicates that codified information, that which was written down, was regarded quite differently in the context of what could be shared and what could not. As a stakeholder noted, in competing for external consultancy contracts, the Greenhouse Accounting CRC had:

> come off an unequal footing against the State agencies in that the State agencies have fundamental ownership over the IP [intellectual property] in the databases they can trade with us in terms of further work and information sharing. The CRC doesn't necessarily have access to that, or unfettered access to pass that on.

So while the immediate relationship between the contracted parties was paramount in this model, other relationships beyond the immediate ones were also affected.

As noted in Chapter 2, the consequences of this 'knowledge economy' perspective on research were ambiguous. Several researchers perceived the purchaser–provider model as a threat to the quality of the science, especially when the results are delivered in a more abstract form, such as a computer model. As one researcher described it:

> The problem is when you're into delivery and you're using models to deliver —you have to make decisions, so you can't afford to walk away saying 'we don't know'. You do have to write an equation, and you do have to put coefficients on the equation and you do have to run it because saying that we don't know is not acceptable.

In more extreme instances this led to a kind of identity crisis among some researchers, who felt real conflict between the demands of the purchaser–provider framework and their own self-image as a scientist. Some researchers felt pressured to shape their 'products' to suit the purchaser, sometimes against their better scientific judgement:

> The trap is, we could easily produce some numbers that are just garbage, and there's a big push for any number which I think would be more damaging than saying we don't know. But there's big policy demand for some numbers in black and white, and they'll settle on anything almost. Provided you've got a number in black and white—and they don't want any caveats or error terms. They'll give you a lot of money if you want to do it, but it's selling your scientific soul in some ways.

In other words, the assumption of the transferability of scientific information was flawed, as the contractual arrangement generated pressure to conform to the expectations of the purchaser.

In many respects, the contracts operate as a filter between the purchaser and the provider, or, in terms of process, perform what Bowker and Star (1999) refer to as 'clearance'. Clearance is a type of organisational forgetting, where historical circumstances and contingencies are swept aside, and processes start anew. The contract serves this purpose of clearance, as by dictating what the purchaser wants to know it simultaneously dictates what they do *not* want to know. No error terms, no uncertainties, no caveats.

By clearing away the scientific past and asking for selective repackaging and re-presentation of research work, the purchaser–provider model requires substantial transformation of information as it leaves the realm of science and enters the world of 'management'. The uncertainties and vagaries are left behind in the science world, and only the certainties—or the appearance of certainties—are transported to the purchaser's world. This is not simply a positivist notion that there is a real world that science alone has access to; clearance is not merely an inappropriate abstraction for the sake of convenience. Rather it is a strategy that "provides a way of managing a past that threatens to grow out of control" (Bowker and Star, 1999, p. 263). In this case, adding the complexity and uncertainty of science to an already highly complex and uncertain political and managerial world could well lead to an 'out of control' situation, resulting in complete inability to justify any course of action.

Embedding the relationship between the researchers and the end users in the legal-administrative system had also particular effects on the ability of the CRC researchers to work together collaboratively on non-contract work. Where different groups within the CRC were competing for funds outside the CRC, the effect on people's willingness and ability to integrate information was hampered as information must be protected to maintain a competitive advantage. The consequences of this on the integration outside the contracted work itself can be substantial:

> it limits collaboration, information and data exchange, knowledge... the competition for funding from external agencies is based on one's ability to do the job, provide some innovative or creative breakthrough, and if you have any leakage of information through to your competitors, then you're going to lose out.

The 'leakage' of information then outside the small exchange system that is created in the purchaser–provider model was potentially harmful and therefore to be avoided, limiting the extent to which people exchanged information beyond the parties directly contracted.

On the other hand, while cooperation outside the project may be hampered, cooperation inside the project could be enhanced, as the pressure of meeting contracted deadlines and milestones acted as a strong incentive for researchers to put their differences aside and just get on with it. One project in the Coastal CRC was generally considered to be highly successful, bringing together people from two universities and several government agencies with disparate disciplinary backgrounds, and local government representatives, for a particular contract. The project leader believed that the pressures of the contract in this case acted to bring people together quickly and efficiently:

> because we had to deliver, everybody said 'right, head down bum up' and got on with it.... I can see the others that haven't got such strict timelines that everyone's still off doing their own thing, there's not that sense of urgency....

In other words, the contract overrode the other differences and served as a firm basis for integrated research.

In summary then, the purchaser–provider model framed integration as a transaction, supported by rational, immutable legal and administrative systems. It honoured the economic-rational construct of consumer sovereignty, in that the purchaser as the contracting party retained most of the control over the research product, if not the process. Purchaser control allowed them to specify what was excluded from the contract as well as what was included. This served a function of 'clearance', so that the scientific input decreased, rather than increased, uncertainty in managerial or political decision-making. While the CRCs as research providers could choose not to take up a contract, the CRC Program administration encouraged them to adopt this mode of research.

Model 3. Jigsaw

In the jigsaw model, people saw integration primarily as an issue of coverage: the purpose of integrating was to bring as many pieces of the information jigsaw together as possible, to create a complete picture. The jigsaw model was an heuristic used to capture a sense of integration as countering the fragmentation of data and knowledge inherent in disciplinary and institutional structures. It was based on the view that such fragmentation was a significant reason why science was not more widely used in the world of action. As one researcher described it, policy-makers need 'structured' output, not just the pieces:

> *scientists will provide a whole bunch of pieces of the jigsaw, because they look at things in a reductionist way, but not necessarily how to put them together in a structured output. Policy people, they need to pick it up so that they can develop policies. And I think that's probably changing as we have more integrated science in the CRC, among other institutions.*

This is illustrated in Figure 16.

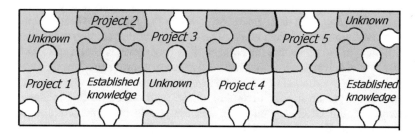

Figure 16. The jigsaw model of integration.

The jigsaw was also an investigation of gaps and overlaps: without cataloguing what was already known, it was very difficult to assess what research was still needed, and even harder to strategically prioritise and plan what needed to be done. Some of the pieces may already be well-established scientific knowledge, others may be work in progress, and still others may be unknown. Integration was about bringing these together.

The jigsaw was a significant metaphor, as it indicated an understanding of the fragments of science as potentially dovetailing neatly in with each other. While there may be some allowances for gaps or overlaps, the overall aim was to have a mosaic, a complete picture of 'what is going on'. This completeness was seen by several participants as being both desirable and possible:

I think when teams realise that there has to be a whole network, like a jigsaw puzzle of information before the picture is fully understood, then you can make better decisions.

In other words, the fragmentation of science (that integration is trying to overcome) is the product of a rational, mechanistic process that has led to detailed knowledge of smaller parts of the world. These can be rebuilt into a detailed, meaningful understanding of the larger world, and the missing pieces, as in a jigsaw, are aberrations, or oversights, where the systematic processes have failed.

Ideally, the pieces do not need to be reshaped to fit—the jigsaw metaphor indicates a belief that the pieces of information should neatly slot together. The appeal of this perspective to people trained to think of research as a rational, logical process of deconstruction was strong: the jigsaw is simply a reversal of a linear, mechanistic process to rebuild what was deconstructed. However, this ideal was rarely realised, so negotiation between those who generated the information pieces and the integrators to provide the 'shapes' required tends to be a part of this model.

This model is overwhelmingly about knowledge and information, and how it fits together. Human relationships are largely the product of the part of the world their information relates to—ecological information goes 'here', biochemical information goes 'here', and so on. Some negotiation with their 'knowledge neighbours', those with knowledge that is juxtaposed with theirs may be necessary to work out any gaps or overlaps, but these interactions, under this model, are solely concerned with the resolution of technical issues. The information that comprises the pieces is disembodied, in the sense that the context of its production is irrelevant, and hence the people themselves are mostly peripheral to the integrative process.

While this is a simplistic presentation of the model, and the way relationships are constructed within it, it is a tempting model with respect to CRCs. One of the advantages of the CRC structure was that the people involved often covered a significant proportion of the Australian scientific community actively working in the particular problem area of their Centre and were often aware of (if not directly involved in) related work outside the CRC. Consequently CRCs were a unique opportunity to complete a jigsaw—indeed, they were almost designed to be viewed in that way. Many of the scientific pieces were captured within the CRC, and they were possibly the best option to build the most complete picture.

The consequences of this model are clear. As indicated, problems tended to be represented as aberrations from the rational ideal—missing pieces, mis-shaped pieces that required honing and reshaping, overlapping pieces, and so on. In some cases pieces needed to be created from scratch. This was the rationale behind the Greenhouse Accounting CRC's inclusion of a program that consisted solely of basic biological science, even though CRCs were generally concerned with more applied research. The argument ran that there were enough gaps of significant scale to warrant the basic research program—otherwise it would be easy to just focus on the pieces we do have, and overlook the ones that are missing:

> *without the reductionist side of things... we would get too confident in our pragmatism, and I would say that's probably the danger of organisations like... ourselves, is that without any reductionist science going on, or process science, or basic science or whatever—we won't actually be confronted by those surprises, and reminded of our ignorance.*

It was also easy to assume that people knew what the picture was that they were trying to construct, an assumption that some stakeholders rejected. For example, one senior bureaucrat noted a general inability of scientists in the Greenhouse Accounting CRC, and the scientific community more generally, to see the picture that was relevant to him:

> *I think that even today, three years on from Kyoto, there are actually very few of the scientist community who actually have an understanding of what the overall picture of the technical dimensions of carbon accounting looks like, and what it means for a research agenda... the way all the pieces of the jigsaw fit together, what the requirements of the future are.... if you're struggling to understand how the picture looks you are struggling to put the pieces together, to put together the integrated research agenda that is needed.*

In other words, the pictures being aspired to by the researchers were inadequate if they did not take the overarching policy context into account. It was certainly possible for some of the puzzle pieces to be non-scientific information; however the degree to which non-scientific information was included in the puzzle varied.

Questions were also raised about who holds the pieces, and different CRCs had different perspectives on this. The Greenhouse Accounting CRC tended to view the pieces as scientific—even the title of their annual get-together, the "Annual Science Meeting", emphasised this. In contrast, the Coastal CRC saw stakeholders as holding important components of the puzzle, a view that was acknowledged by stakeholders:

> *the Coastal Zone CRC is probably the best research body I've worked with in terms of trying to come to grips with integration, community participation. Not to say that they've got it right, but they are a CRC who has a clear focus that that is one of the important things....*

In summary then, the jigsaw model of integration was a popular heuristic that implied a technical, mechanistic approach to integration, which reversed the rational, logical process of scientific reduction. The result would be a full range of pieces slotted in neatly together to create picture of complete understanding, which would be relevant to non-scientists. While the scientific information represented in the CRC clearly fit the role of the pieces, the role of non-scientific information in the puzzle—as well as who the final picture was actually for—was ambiguous.

Model 4. Silos

The silos model of integration was about actively *combining* different types of information and knowledge. Like the jigsaw model, it was also about bringing together what had been fragmented, but it wasn't enough to fit the pieces together: the information had to be mixed, reformulated or repackaged before it was passed on to users.

The silos model was manifested mostly in the organisational structures of the CRCs. As mentioned in Chapter 3, both CRCs had an integrative centrepiece: in the Greenhouse Accounting CRC an integrative model that was intended to draw together a wide range of the CRC's research into a single model; and in the Coastal CRC a multiple objective decision support system, again intended to incorporate the scientific information generated throughout the CRC with stakeholder-relevant information to improve decision-making. Similarly, the Coastal CRC also developed an integrated modelling project, which was just starting as this study concluded. This project was to serve a similar function as the Greenhouse Accounting CRC's modelling centrepiece. The use of a modelling technique to integrate different forms of information is represented in Figure 17.

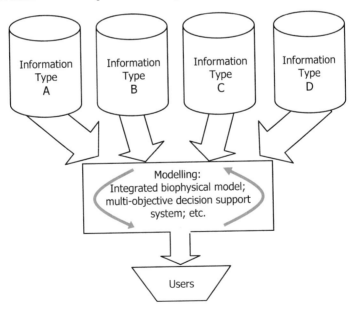

Figure 17. The silos model of integration.

Many researchers referred to the processes of 'feeding' from one project to another, as a key to this model. For example:

It will be interesting to see by the next three years how the projects evolve, how they do feed into one another, and whether we have been able to foster greater integration.

and:

I could see that if we ever do run a carbon model, everything I do basically will feed into it.

Consequently the importance of the silos in this case was not the separate information types themselves, but that the information was 'poured' into an integrative process.

As for the jigsaw model, in the silos model the connections between people were shaped by their participation in a technical process. This process was even more technically driven than the jigsaw, as the information was passed on to others to manipulate, rather than being negotiated at the edges. Again the information is disembodied and a-political, in the sense that it is presumed that it *can* be meaningfully combined through these abstract processes, a presumption that one participant noted was not shared 'outside' science:

...there's a lot of uncertainty in government agencies about using modelling to help decision-making. It is usually very much political processes that drive policy, it's hard to see someone in the EPA [Environment Protection Authority] working over a few months to make a decision using a more abstract process in modelling.

Having said that, in both cases the models remained scientifically driven, in the sense that the researchers managed the way the information was brought together and given meaning. While it was too soon to tell how effective they would be in this task, as their implementation was to be ramped up over the life of the CRCs, early indications tended to be sceptical rather than favourable. In other words, the 'solutions' proposed by the models, while rational and scientific, were entering a different political realm when they were put to use by non-scientists. In this other realm, the rationalities in use may be quite different.

In the Greenhouse Accounting CRC the integrated modelling project was also perceived to be controversial. This controversy lay within the science community, rather than with external stakeholders—many scientists were not comfortable with having to shape their research design or outputs to suit a modelling project. This may be because information providers had little say in what happened to 'their' data or input once was integrated into the model. Indeed, in some respects the simplicity of the silos model served to highlight the opacity of these integrative processes, through the 'black-boxing' of the modelling or other technical process. Researchers and end users often felt that they did not have much say in how the information they generated was going to be used in either the modelling or the eventual decision-making. This opacity generated some cynicism:

And then, of course, there's the 'integrated modelling project'. Which certainly initially had the aim of taking the results from the individual projects and integrating them into some magic model that was somehow going to be able to operate at the projects to regional and national scale.

So, although the silos model of integration was relatively simple in structure, many people recognised that the processes embedded within that structure remained complex, uncertain, and largely opaque.

The view of integration here is technically sophisticated—the modelling processes tended to involve a wide range of ways to approach diverse information sources (databases, large or small, process understanding) and to capitalise on modelling work already done without reinventing the wheel. In other words, the integrative processes here were often 'meta-models' in the sense that they were attempting to incorporate existing models rather than build a new model from raw data. They tended to be run through complex computer programs that were accessible only by those trained in their operation.

Yet while these technical processes gave a sense of objectivity, they were, of course, political themselves. For example, who was included and excluded from the integrative relationships varied according to the models. While the Greenhouse Accounting CRC's Integrated Modelling Project dealt exclusively with scientific information, where the 'Information Types' of Figure 17 were different disciplines or even existing models, in the Coastal CRC's Multi-Objective Decision Support System, these information types included scientific, community, and government information, and were flexible to include a range of stakeholders who wished to participate. The model, in this instance, provided a technical process through which these parties could come together and work through complex scenarios. These can be understood as strategic decisions—it was important that the Greenhouse Accounting CRC, given their political context, retain the appearance of independence from their stakeholders. One researcher expressed this clearly:

The challenge with AGO of course is that we have to remain both cooperative with them, because they have a mandate to run on a lot of these issues, but also be sufficiently independent of them that ... a CRC statement which comes out and supports the Australian position is seen to be independent research, not as a spokesperson for the AGO.

Similarly, the Coastal CRC highlighted their Multi-Objective Decision Support System process, as it explicitly included stakeholder input, a key part of their mandate to 'bridge the gaps' between science and policy. Indeed, the prominence of these models in both CRCs illustrated the rationale behind the silos: to counteract that many scientists within the CRCs were still carrying out traditional reductionist science, the CRCs needed to demonstrate that this work was going to be actively integrated. As such, many more isolated projects were intended to 'feed in' to the integrative ones.

Once again, this process of transfer from one context to another involved removing a part of the context of the information being fed into the models. As one modeller acknowledged, many researchers may find the extent of this removal—what Bowker and Star (1999) have called erasure—alarming:

> *I think for the most part, people in themes five and four are probably wondering just what is going to happen in Decision Frameworks. And ... I guess they might be quite alarmed to some extent as to how little, or perhaps how much, of their work will actually go all the way through to decision-making and policy formulation.*

Indeed, the control over this process had considerable political ramifications within the research community, which will be discussed in more detail in Chapter 5. The different approaches to integrative modelling illustrated by the two CRCs, in particular the different roles granted non-scientists, entailed different processes of erasure. In the Coastal CRC, the MODSS process gave the non-science 'users' a significant say in what was to be included and what was to be erased. In the Greenhouse Accounting CRC the researchers, in particular the modellers, retained control over these processes.

To summarise the silos model, it was primarily a technically driven approach, with integrative modelling tools as centrepieces towards which all other research in the CRC fed their results. These technical processes 'integrated' the information that was delivered to them, and in turn delivered transformed, packaged information (or processes) to stakeholders. This model allowed each of the CRCs to highlight an integrative centrepiece that aimed to bring their disparate research programs together.

Model 5. Value adding

The value adding model was similar to the silos model, but with a multi-stage process of incorporating information between research projects and programs, rather than tumbling them all into a single integrative black box. The Coastal CRC's triangular program structure, illustrated earlier in Chapter 3 (Figure 9), demonstrated this approach. The value adding model of integration resembled a production chain, where at each stage of the process further manipulation and development added to the value of the product. It was more complex than the previous models, as it had links built in across all programs, rather than a single integrating project as in the silos model. Responsibility for integrating research results was therefore shared across the CRC, rather than being the sole domain of the integrative project. As Figure 18 below illustrates, there are 'integrated' applications at each of the different levels, representing different scales of integration.

In this case, the results or outcomes from the first layer of research act as building blocks that feed into the next layer up, and so on. The final level uses frameworks that encompass all of these types of information, to: "integrate social, economic and ecological knowledge with stakeholder involvement as a basis for governance for future sustainability in the coastal zone" (Coastal CRC, 2000, p. 10).

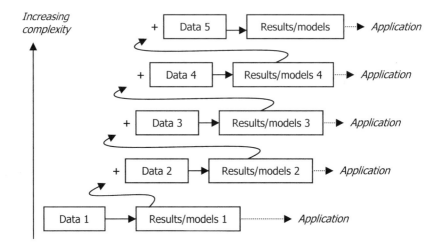

Figure 18. Value adding model of integration.

The value adding model actively incorporated many of the CRC Program goals, actively integrating science at many scales, and delivering the outcomes to a range of stakeholders at different levels of complexity. When the model was shown to researchers who had been unfamiliar with it, several commented that it was ambitious (perhaps overly so), but also that it was possibly 'the ultimate' model for CRCs.

Indeed, the expectations of what could be achieved through this approach were high:

> *There were very high expectations raised, with CSIRO and the great integrated computerisation, remote sensing, all those things. We have to meet those expectations.*

Consequently the value adding model was, like the silos model, concerned with bringing together scientific information in ways that provided a more complete picture for non-scientists to work with in their decision-making processes. It was also similar in that it was a highly structured research model, where the information produced by each individual project was part of a grand design. While individual projects within these levels could and did include stakeholders in their research, the value adding process was primarily about linking scientific (including social science) results, to once again build a more complete picture:

> *I think—not components or discipline science, but integrated pictures. People can say 'I'm really working on these mud crabs at Yeppoon' but how it fits within the food web, or how it fits within the structure of this bigger picture. So that people can be team players in advancing the whole coastal ecosystem before them. And as part of that they can recognise they have got their own important key skills in that component but there are other people that they need to work with. To go forward...*

However, the difference in this model was the adoption of an explicit structure for different types of information to feed into, based around the concept of increasing system (and management) complexity. As indicated in Figure 18, this did not necessarily mean that the research 'became useful' at the end of the process, but that each layer had different application contexts, depending on the level of complexity of the decision-makers' demands. For example, a community group wishing to aid the preservation of the local mud crab species is likely to need the species level data as well as, perhaps, ecosystem level data and outcomes, whereas the local city council may wish to know the social, ecological and economic impact of a new sewage treatment plant, drawing from the top of the diagram.

While the arrows of Figure 18 indicate the primary flow of information, there was also a sense expressed by participants that people were not anonymous information suppliers or consumers. Like a sensitive production to consumption chain, the demands of the variety of users/consumers led to a variety of different 'products' and interfaces between information consumers and suppliers.

The 'consumers' were of two quite different kinds: there were the 'scientist' consumers at the next level up the scale of complexity, indicated in Figure 18 by the solid line between one level and the next; and the 'non-scientist' consumers, indicated in Figure 18 by the dotted line leading out to 'application'.

This approach encouraged small-scale as well as grand-scale integration, as there were more opportunities to work across the organisational boundaries set by the CRCs. However, they also needed to be compatible with the grand design—or *vice versa*. As noted in Chapter 3, the Shavian Triangle was promoted quite heavily both within the CRC and externally, as central to the CRC's identity. This strong commitment to the model, aside from creating high expectations, also rendered it quite inflexible. This also limited the Coastal CRC's ability to alter the research structure built on this model as the CRC developed, or as people found other structures, such as the Management Study Areas, more conducive to how they saw their work in relation to others. As a result, several researchers viewed the Shavian Triangle as little more that a public relations exercise, useful for 'selling' what the CRC did, but not directly relevant to them.

In summary then, the value adding model for CRCs was a complex, highly structured model in which a range of research projects worked together in different 'layers', to feed information to the next layer to integrate information of ever-increasing complexity. Researchers were expected to work closely together, and with stakeholders, so that relevant integrated outcomes could be generated at each level. In the degree and explicitness of integration, this model was regarded by some as the ultimate CRC research model, albeit complex and difficult to implement.

Model 6. Extension

The extension model reflects some of the developments discussed in Chapter 2. It was primarily concerned with relationships between the research and non-research groups, often referred to as 'end users' or 'stakeholders'. In this model, integration was seen to be educative, allowing and encouraging non-science groups to incorporate scientific information into their decisions and activities. 'Getting results out there' was seen as a key to integration. The language of the extension model was varied, but usually reflected the constructs of the 'transfer of technology' concept discussed in Chapter 2. Talk of transfer and translation were key indicators that this conceptual structure was in use. For example, one researcher described a project as:

> *looking at environmental planning and how the concept of sustainability can be transferred to local government planning schemes in a consistent and integrated way.*

while another spoke of transferability:

> *I think in terms of outcomes. Because [the CRC's] involved with state government agencies, local councils, a whole lot of stakeholders, it's very easy to transfer recommendations.... I see that as very good in terms of transferability of the research to management.*

In either case, the primary flow of information was from science to decision-makers, whether directly as relevant information pieces, or 'pre-integrated' through scientific tools such as models or decision support systems, as illustrated in Figure 19.

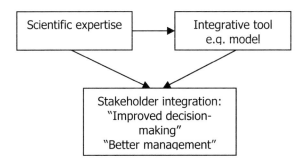

Figure 19. Extension model of integration.

This model was implicit in several previous models, as the point of connection between researchers and 'stakeholders' or 'end users', but it was also used more generically. Some viewed the integration achieved through extension as only one aspect of a far more complex decision-making scenario:

> *there are always multiple objectives in natural resource management, and there are competing influences and there is always going to be a range of social and economic and political and scientific information that... is a basis for decision-making... I guess the integration is how well, and how early that information comes together for decision-makers to consider all aspects of an issue....*

However, the dominant view of integration as it related to extension was that it was the researchers' role to bring the information together *before* it is presented to the decision-makers:

> *I guess the integration comes from being able to draw on experiences or data or the use of models, and apply them under newer scenarios than it was earlier, or originally used.... to either develop scenarios or to evaluate options and then take those outcomes out into environmental planning, or policy and development.*

This links the extension model with other models concerned with how the research comes together into an integrated product.

Despite differences in perceptions of how subtle the relationships between the scientists and the non-scientists were, adoption of the extension model of integration inevitably differentiated the participants in the model (the researchers and the non-science others) in juxtaposition as experts and non-experts. In caricature, the key assumption of this model was that science was a source of expert knowledge, and that knowledge needed to be adapted for use by the non-expert.

Information products were also prevalent in this model, which were again information, 'packaged' or 'translated' to be suitable for non-scientific use. One researcher described his preferred role explicitly in those terms:

> *we should be packaging—both driving the agenda and allowing information to be packaged in a way that it is actually useful in the policy setting.... If we can educate the policy people to be ahead of the game in both our data and analytical capability, so that we can say what's feasible and what's not from a policy point of view. Yeah, what policy options are or aren't feasible from a scientific point of view.*

Similarly, a communication manager expressed the belief that all researchers and scientists should take on the task of extension and translation:

> *I mean everyone has the role of communication coordinator expected as... part of their job. And similarly as an external extension agent I guess, to help translate the results into useful actions or useful policies or useful practices.*

Consequently, as for the purchaser–provider model, it was premised on the belief that scientific information was relevant to non-scientists, but that those non-scientists were a broader category than those with a financial investment in the research.

This accentuation of the authority of science was an inevitable source of tension in the CRCs as, first, the distinction between expert and non-expert was often not clear cut. This was especially evident in the case of the Greenhouse Accounting CRC, where stakeholders such as the Australian Greenhouse Office—a 'non-research' partner—were staffed by people with considerable expertise in greenhouse science, notwithstanding their greater expertise in greenhouse policy, an equally specialised area of expertise. Also in the Coastal CRC, business, government and community stakeholders often had significant scientific training as well as detailed local knowledge of both ecosystems and management systems. Secondly, in some instances, both researchers and stakeholders actively resisted the characterisation of expert versus lay-person, which placed them in conflict with others who unquestioningly accepted the authority of science. This was particularly evident in the Coastal CRC, as their efforts to include stakeholders had attracted researchers who valued non-scientific knowledge and did not see themselves as experts in relation to non-scientists. These researchers tended to be critical of the Coastal CRC's claims to be 'participatory', for example:

> *[Other researchers] don't seem to understand or value anything outside of the science, they seem to equate science with research and think that the only valid form of knowledge is science. And that's really incompatible with being inclusive of non-scientists. Those sorts of problems haven't really been addressed.*

This indicated a gap between the Coastal CRC's preferred approach to move away from the classic extension, transfer of technology model and the capacities of many researchers to 'do participatory science'. Indeed, an internal study found that of the many projects proposed to the CRC that made claims or expressed wishes to be participatory, only 43% had incorporated this into their actual research methods. (A follow up study, conducted a the end of the CRC's life in 2006, found that this figure had increased to 62%).

In contrast, the Greenhouse Accounting CRC made no claims to be participatory, and did rely more on conventional extension and transfer mechanisms. This may reflect the different policy contexts of the two CRCs, and their consequent source of authority. As the Greenhouse Accounting CRC was positioned within a volatile political arena, their authority depended primarily upon their independence and good standing as scientists. As one member noted, there was a fine balance between being a political player and being a rigorous scientist, referring to the difficult job:

> *of balancing out the needs of providing robust, rigorous scientific advice and possibly being captured by the negotiating process to the extent that you make the facts fit the party line.*

Their approach to modelling reflected this view, in that the centrepiece integrative model was solely concerned with bringing together biophysical information from within the CRC, with no formal avenues for non-scientific input. In this way, the researchers retained control over the processes of erasure entailed in simplification.

However, others within the Greenhouse Accounting CRC were very aware of the fallibility of the science they were producing, and thus were cautious of setting themselves up as experts. For some researchers this was exacerbated by the extensive reliance on large-scale models. As one researcher noted:

> *A lot more time I think we should be saying we don't really know and we shouldn't be making decisions—or you go back to your policy based*

decision-making where you deal with a hundred percent uncertainty and you make decisions on some other basis rather than the science numbers.

The extension model encouraged researchers to consider the relationships between their research and its eventual outcomes in action, but it also encouraged them to imagine themselves as the experts within those relationships. The emphasis within this model on 'transfer' also suggested that it was desirable and possible to ship information from one to another without it being transformed in the process.

However, at the same time researchers were often aware that the 'scientific' outputs were not what non-researchers wanted or needed. Considerable detail needed to be erased so that significant details were noticed. Communication managers were often in the midst of this delicate process, assisting and encouraging researchers to participate in extension activities. They were aware of the complexity of the tasks—as one remarked:

when [researchers] get to the stage of putting as much effort into planning [communication] activities as they do into planning the research design, then I think we will have really succeeded.

Consequently the 'packaging' of research, although usually spoken of in a slightly disparaging way as though it were a non-scientific pursuit, actually involved the careful balancing of scientific relevance with enough erasure to avoid overwhelming audiences, while at the same time serving the politically and socially sensitive task of interaction across science–non-science boundaries.

In summary, the extension model was concerned with how researchers 'translated' and 'packaged' their research for external consumption. While this model relied on the authority of science as a producer of knowledge, the political contexts of the two cases had a substantial impact on how this could be integrated with non-scientific knowledge. In each case the authority of science remained, but the extent to which it was actively integrated by the researchers into decision-making processes outside science differed.

SOME COMMONALITIES

While these models illustrate the differences between ways people made sense of integration as an abstract concept, there were several important similarities between them.

Rationale for integration

In most cases the primary rationale for integration was the recognition that research, and hence the knowledge and information it generated, was: isolated (in the sense that people did not know what others were doing); fragmented (what they were doing did not dovetail neatly with what others were doing); dispersed (physically, leading to fragmentation due to differences in research environments); diverse (in an unbounded universe of potential research projects, those selected were quite different); and individual (in that research tended to be carried out by individuals or small teams that were 'owned' by those individuals or small teams).

Primacy of information flows

While there are many potential similarities and differences between these six models, one that is particularly important is that they are all *primarily* focused on information/knowledge flows, with the models themselves representing how that flow was structured. It is information that flows from the silos to be mixed, people being juxtaposed were seen to represent bits of information, information was the raw material in the value adding and silos models and information was being purchased or supplied in the purchaser–provider model and the extension model. Further, the information that was relevant in these formalised conceptions was overwhelmingly scientific or technical; even when the participants themselves were not scientists, they contributed to the integrative process by articulating their demand for scientific research, or by agreeing to participate in a technical–scientific process.

Accumulation, manipulation and presentation

In all models bar the first, integration as a goal was inherently ambiguous—perhaps even confounded. As a process of bringing together a wide range of scientific information, integrative tools operated at a high scale of complexity. The information was multi-disciplinary, sometimes it had already been modelled, it was invariably originally designed for a different purpose, and there were gaps and overlaps to be dealt with. Yet the purpose of integration was often (although not exclusively) perceived to be the provision of simplified information outputs to stakeholders.

Despite their differences, this process of information flow was central to each of these models: the information is first accumulated, then manipulated, and finally presented 'out' to the wider world. So the integrators had to be able to understand both the massive complexity of the scientific (and sometimes non-scientific) information coming into the integrative processes, how to manipulate that, often through sophisticated technologies; and how to simplify or "re-present" it so that it remained suitable for non-scientist consumption.

While presented in the positive form as representing increasing knowledge, these processes may be viewed comfortably from the perspective of integrated research being an extension of traditional objective, additive or accumulative, independent science. However, when viewed in the obverse, integration is simultaneously a process of exclusion (which information is not included), erasure (which information is manipulated out through processes of simplification) and clearance (what gets left behind in the science world when the information is presented elsewhere). In considering both sides of integration, knowledge is not necessarily being increased, or necessarily decreased, but being re-presented via processes bound within socio-political, cultural and technologically-driven contexts.

This means that processes of integration are inevitably processes of substantial erasure or even clearance, of stripping out details and seeking essences according to particular application goals and contexts. It is a process of *decontextualisation*. As indicated in a preliminary fashion in this chapter, this decontextualisation— even while it aims to reduce the personal aspects of the research even further—is inevitably a social and political activity, as it involves manipulating the work of one group to meet another group's perceived needs. While the constructs of integration as illustrated in this chapter were primarily concerned with information flows that *presumed the information was already decontextualised*, the practice of integrating was deeply concerned with the social and political dimensions of working within these borderlands. This is the subject of the next chapter.

CHAPTER 5

EXPERIENCES OF INTEGRATING

The models discussed in Chapter 4 show that people overwhelmingly thought of integration as a process of managing and manipulating information flows. The flows were integrated through various designs, and relied on the information being representative, rational, and above all, impersonal. Human processes were significant in these models in relation to the transfer of information, its technical integration, and packaging for various audiences.

This chapter shifts perspective from questions of *what* integration was thought to be in the abstract, to questions of how it was achieved in practice. How did information come together as these models of integration were implemented? The models of Chapter 4 suggest that technical issues, working out the compatibility of different information types, analysis of gaps and overlaps, and similar biophysical concerns would be the main topics in the practical implementation of integration. By categorising integration as a predominantly technical issue, it follows that action—the implementation of the models—should likewise be technical. Yet questions of 'how', while sometimes concerned with technical issues, were more commonly focused on the social and personal dimensions of integrating. In theoretical terms, the activity of integrating—as opposed to the abstract constructs of integration—was thoroughly contextualised.

FROM PRACTICE TO PRESENTATION

While the links between the 'data' of research conversations and the models presented in Chapter 4 were based quite specifically on the ways in which participants construed the words 'integration', the links between practice and presentation in this chapter are more diffuse. In this chapter I draw on the ways in which people described the *activity of participating* in integrated research.

To do this, my analysis turned away from abstract accounts of integration toward close examination of the ways in which participants *contextualised* their accounts of research process and activity. While the notion of context is somewhat problematic (Lave, 1993), in this sense, 'contextualised' activity is activity that is understood *in relation to* historical, social and cultural phenomena. In other words, I examined how participants interwove their personal histories, relationships and broader understandings of socio-political circumstances into their accounts, and *how these were attributed significance* with respect to their participation in integrated research. These ways of attributing significance were again embedded in different categorical schemas, suggesting a different, more personal aspect of learning from that reported in Chapter 4. Once again, I will present fragments of conversations that are grouped together according to the conceptual schemas within which participants made sense of the relationship between their context and practice of integration. While I have tried to present the categories using the language of participants as closely as possible, occasionally I have created new categories to group points that were thematically similar but expressed by participants in different ways. The use of participants' language is indicated by the passages quoted, and by the use of quotation marks.

The ways in which people attributed significance to the historical, personal and socio-political context can be loosely grouped into two categories: working with others; and a personal sense of identity.

WORKING WITH OTHERS

The practice of integrated research was most commonly phrased in terms of the development of relationships over time. To avoid having to pull apart different aspects of experience ('social factors' versus 'historical factors' versus 'political factors' *etc.*) that were closely intertwined, I will use this temporal development to theme the following account. These themes are illustrated in Figure 20.

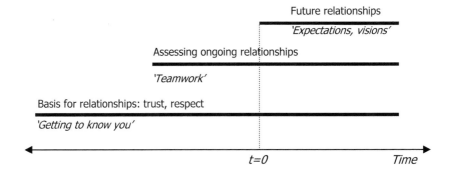

Figure 20. Layers of relationships over time

The first theme discussed here is 'getting to know you', a crucial issue in the early development of the CRC. Following that are discussions of teamwork—when it works, when it does not. Both were concerned with inter-personal, social connections as well as political influences and broader cultural (especially institutional) biases and pre-conceptions. Significantly, the relevant temporal dimension of context did not stop at the present: equally important (occasionally more important) were expectations about the future. The final theme is based around the creation, development and management of expectations.

"Getting to know you..."

As noted in Chapter 3, one of the common motivations for researchers to join a CRC was the opportunity to work with people that they would not have been able to work with otherwise. While it was common for people to know each other slightly, or to know of each other by reputation (as one participant noted "Australia is a very small science community"), many participants spoke of the importance of 'getting to know' each other in a more intimate, inter-personal, processual sense. As one researcher described it:

> *It evolves, it doesn't happen instantaneously. It's just time getting to know people....*

In the early stages of the Coastal CRC's life, efforts were made to encourage this process that went beyond the usual discussions about science. At their first annual meeting, the Coastal CRC organised a number of team-building exercises. The value of this was appreciated by some, suggesting an acknowledgement that 'getting to know people' went further than finding out what they were working on, and whether they were technically compatible, for example:

> *we did a lot of team building. I call it hand-holding. ... I thought that was very appropriate and I was very supportive of that, let's get to know the people we are going to spend time with.*

However, others were not as comfortable with moving beyond the bounds of science, where 'getting to know you' meant getting to know you in the context of your work:

> *the only useful thing [about the workshop] was that we started to get to know other people, but as soon as we started to get down to the nitty gritty we'd have to go and do another [scoff] team-building exercise or something, passing people through strings etc. And really what you need is three or four days sitting around getting to know people.*

The "nitty gritty" was the detail of work and interests that might lead to productive collaboration. This particular researcher did not deny the importance of human dimensions in collaborative relationships—in other parts of the interview this was stressed. However, he did believe that the best way to get to know people was through discussing their science.

Some indicated that this process of getting to know one another was even more important in the context of integrated research than in conventional research teams.

> *Every organisation goes through that period where you are still getting to know each other in the organisation, and I think integration is so much dependent on informal structures and personal relationships, and trust that's built up over a period of time....*

The generality of 'getting to know you' covered a broad, and often fairly indistinct territory. However, as the final quote above showed, some people offered more specific issues as important aspects of this broader inter-personal sense of comfort and well-being. Common themes during the 'getting to know you' phase were the gaining (and losing) of trust and respect.

Trust   Trust was a recurring theme throughout the interviews. The need for trust among scientists was particularly prevalent in the Greenhouse Accounting CRC—ironically its visibility was most commonly a reflection of its absence rather than its presence. Trust was most often regarded as a prerequisite for being able to work together. For example:

> *Building trust is also a big thing in this. I mean there are, realistically, competing groups in the CRC, so building trust is difficult.*

and:

> *integration success is based on alliances of people which is based on trust and good will.*

However, it was also a significant issue between the researchers and their stakeholders, and this arose more in the Coastal CRC than in the Greenhouse Accounting CRC. Several stakeholders emphasised the role of trust in their relationships with the CRCs, and indicated that this trust was not unconditional:

> *It took us quite a while to get that [liaison] person but once we did [the project] moved very quickly, because we could build up the trust in the relationship with the CRC through that person. Up until then we weren't developing any relationship, yet we were willing to put up dollars and try to work something through.*

Likewise, the people in between the researchers and stakeholders, such as communication managers and officers, were also aware that they needed the trust of the scientists "not to misrepresent their science". In other words, trust was seen to be an essential support for the entire network of people involved in the CRC.

The historicity of judgements regarding whether a person could be trusted was often apparent in conversations, with previous experience of a relationship being the best guide to how relationships were expected to unfold into the future:

> *In these cases other people I've worked with in multiple projects, so if somebody comes to us with a question, we think of them as natural partners. So essentially it depends on development of trust and rapport with people. And that's still ongoing with people within the CRC....*

This also worked the other way, where 'bad' experiences in the past made working together in the present and future very difficult. One researcher, commenting on the delays in receiving funding from the CRC noted that past experience was a driving factor in who was prepared to forge ahead without funding:

> *I'm just merrily going along doing the work, but a lot of people who have been burnt by other CRCs who have said that they have actually ended up losing money, they have done a lot of work and never actually been paid, so they think 'why would this CRC be any different?'*

As this comment suggests, in the absence of the inter-personal experience of meeting, working with, getting to know an individual, experience of their organisation, institution or other affiliations were used as a guide. This was illustrated with relationships between CSIRO and the Greenhouse Accounting CRC. Early negotiations between the CRC bid organisers and CSIRO were complex and occasionally acrimonious, and were entangled with the AGO funding for the National Carbon Accounting System (NCAS). Although administratively a separate process, due to the relatively large demands of the NCAS program in comparison with the number of researchers working in carbon accounting-related fields, most of the CRC researchers were also involved (or wished to be) in the NCAS. Consequently, experiences gained in the NCAS process inevitably shaped the perceptions of people in the CRC. As one researcher commented, this situation was not unique:

> many of the other groups in the CRC are also our direct competitors in terms of external funding. Once again this occurs quite often in science. The people you have to compete against, you also have to work closely with. There no rules of course on how you do that, it relies essentially on personalities and trust.

CSIRO played a significant role in NCAS, its participation earning the organisation descriptions of being 'vicious', 'aggressive' and 'predatory'. Loss of trust through this process posed particular challenges for the CRC, both for upper management attempting to develop functional teams of researchers formed by many of those who felt they had been 'screwed over' by their colleagues, and for those researchers themselves, who were expected to work in close collaboration with their competitors.

> For example one of the organisations in the CRC explicitly tried to get our funding. We were asked to leave the room when they did their presentation, and then they made direct criticisms of our approach to the funding agency... [We refuted] the accusations they had made, and showed that their case was false. Now, we have to do business with those groups in the CRC

These sorts of histories meant that few people had the luxury of entering these relationships with 'a clean slate'. Even an institution's or organisation's reputation, for example, affected perceptions of trust. This also influenced relationships beyond the CRC. As many of the CRC partners were large organisations in their own right, such as CSIRO and Federal and State government departments, it was difficult for those outside the CRCs to know exactly *who* from these partners was involved. Consequently, when those outsiders had bad relations with some within those large organisations, it was easy for them to assume that their CRC colleagues were now working their 'enemies', even though the individuals or groups involved may be entirely different:

> we don't actually say the organisation is competing against us in that regard, so we do business with other people in that organisation based on good will and all those things. But our collaborators don't see it necessarily that way. They don't see it as individual personalities for example, people tend to lump things and say 'well it's that mob over there'.

This episode has been recounted here at length because it illustrates the complexity of personal, organisational, historical learning that contributes to the development and maintenance of trust. It was not only the history of the CRC that was taken into account, but the history of the partner organisations, their previous interactions, their reputations, and their cultures that were inextricable in judgements of trust.

Respect   Respect was a highly significant issue, as there are many deeply-held differences that can lead to people being sensitive about whether they are being shown an adequate amount or appropriate type of respect. Respect tended to centre on a person's professional competence, and their personal ethics or values.

Respect for competence was a sensitive issue across the two cases, but it was more prominent in the Coastal CRC than the Greenhouse Accounting CRC. Ironically, this was probably also due to the NCAS process. Despite the lack of trust (which was closely associated with lack of respect for values, as will be discussed further below), the NCAS process had at least served as a type of clearing house, and so many of the CRC participants already knew of each other as high-calibre researchers.

The sensitivity of respect for competence in the Coastal CRC was particularly related to stakeholder involvement. As stakeholder participation was a central component of the Centre's rationale, stakeholders were actively encouraged to work with the Centre. Bringing stakeholders and scientists together in the Coastal CRC highlighted the ambiguous position of the scientist, at once an expert and simply another stakeholder. Key to the success of the research, and indeed the Centre as a whole, was the ability of the researchers and research managers to manage this tension.

In some cases this was done well, for example, one project leader was happily working closely with her stakeholders, and welcomed their input into the project:

> *to go out and strategically get samples from the catchment we need them, we need their knowledge. Without that local knowledge we couldn't do the job quite as well.... [B]ut there's also speaking to them about historical things that have happened there. I don't think they think it's that important. We ask them specific questions, but just in conversations they bring up things that are really important....*

This researcher believed that the stakeholders, in this case a local catchment group, didn't have enough respect for *their own* competence. However, in other cases the tension was seen as a challenge to the authority of science and the value of scientific training. One researcher, for example, resented the priority stakeholder needs were given in the Coastal CRC, likening it to 'a patient telling the doctor what to do'.

This diversity perhaps contributed to stakeholders' uncertainty regarding their own competence in relation to that of the scientists. They were often highly sensitive to expressions of respect which, if poorly expressed, could backfire, leading to the reverse of the intended outcome. For example, at a meeting for one of the CRCs, a group of local community stakeholders were invited to participate in a workshop. They were seated in a semi-circle, facing the researchers and other CRC members. As one stakeholder recalled:

It was like monkeys, we felt like somebody was going to start throwing peanuts. It wasn't a good way—they were trying hard to make us feel really important and special but it just felt patronising. You ... almost felt like you were being treated as a bit dumb, that you needed to be brought up to the level of these brilliant researchers.

It was possible to respect someone for their competence, but not their values or ethics. For example, at the same workshop, another stakeholder was reportedly highly annoyed at what was perceived as lack of respect for community monitoring programs:

our Waterwatch person was just furious at that conference [at] the guy [who] said how the community, you know, you can put them in a tinnie [dinghy] and give them a Secchi [dish] and goodness knows they can actually pull up a bit of water! She found that really, really insulting, to suggest that the data because it was generated by the community it is of less value than if you have qualifications or whatever.

Presumably it was not the scientific competence of the 'guy' in this account being questioned, but the way he was perceived to value (or not value) the community.

There were a few discernible 'indicators' used by stakeholders and researchers to gauge respect. One was the stage at which their interests were included in the research. Some stakeholders noted that their interests were being 'added on', rather than being fully integrated into the planning stage, suggesting ad hoc consideration. As one stakeholder representative said:

I do feel like, in one way, the planning for what was going to happen in the CRC, the research projects and the PhDs and all that... was planned and decided before they asked [us] what we thought was important.

Inclusion in the planning stage was an indicator of genuine respect, and there was considerable caution about researchers paying lip service to the idea of stakeholder involvement, but not really wanting their direct participation.

One researcher in the CRC who had extensive experience in working with stakeholders and community groups was particularly sensitive to this issue. As mentioned briefly in the previous chapter, she was supported by the CRC to analyse the range of projects being carried out across the Centre to see how many had actually incorporated methods to include stakeholders, how many included stakeholder involvement in their goals but not their methods, and how many did not make any claims to stakeholder involvement at all. In her analysis, while 30% of the project proposals incorporated methods to actively involve stakeholders, 20% expressed intent to do so, but without the required methods; 25% claimed aspirations to include stakeholders, but did not build their involvement in to the project; and (presumably) the remaining 25% expressed no plan or aspiration to include stakeholders at all. This was to be used by the CRC as a benchmark enabling them to monitor improvement (or decline) in their stakeholder involvement over time.

Other indicators of respect for stakeholders included travel. One researcher who had carried out a highly successful research program that was stakeholder-funded and supported, noted the importance of physically visiting the people they wanted involved in the project:

> *It wasn't when we brought people into [our] context, it was when we went to them.... I know my way around every south east Queensland council, I can drive you there without a map [laughter].... They want to know two things, they want to know what's going on, and... they want somebody to listen to them. What they have to say.*

Respect was a crucial dimension of the social relationships between CRC members, particularly across institutional boundaries. In these boundaries, scientific competence was rarely sufficient to garner the respect required to build a working relationship.

Trust and respect entwined   While they have been presented largely as separate issues here, trust and respect were usually heavily entwined. In large part the 'getting to know you' phase of the CRC can be viewed as a process of establishing trust and respect, as groundwork that enables the 'real work', the science, to happen. The researchers were mostly aware that this was a complex, sensitive, challenging process, involving personal, organisational and institutional affiliations. As one researcher commented, a vital aspect of success in integrated research was:

> *the ability to downplay the institutional interests versus the interests of the CRC as a whole. This also means the development of mutual respect rather than thinking 'representative of institution A said this work by B and C is really not up to scratch' is really a valid judgement, not an expression of self-interest. And it's only when you've got a working relationship and you trust them that you know that that judgment does has something behind it rather than just manoeuvring for another slab of money.*

This inter-personal groundwork was often viewed as a disadvantage or overhead—a necessary cost, but a cost all the same. This was one of the reasons why some viewed research that was integrated across different research organisations pessimistically. For example:

> *There's a lot of time spent in spinning wheels, a lot of inertia, a lot of effort goes in to just getting people together whereas through the one institution you can feed into an existing culture, the marginal dollar goes further I believe. One extra dollar to my group, and not having to have a separate series of annual reports, a separate series of milestones....*

To follow the economic metaphor then, the establishment of trust and respect might be regarded as a 'sunk cost', something that is necessary for the 'production' of integrated research outputs, but not recoverable.

Others viewed the opportunity to 'get to know' others, both researchers and non-researchers, positively. To them the groundwork was not a sunk cost, as it was a valuable process in its own right:

> *So for us... there is a lot we can gain, and little we can lose. Even if we don't recoup all of the in-kind that we put in—although it looks like we are, quite happily—that's not our principle concern. Our concern is what we can learn from our experience. And I think the local community is the same.*

As the last quotation shows, the difference between these two perspectives may be understood as attitudes towards learning—is what is being learned through participating in the process of doing the integrated research itself of value? In the former scenario above the answer was most likely no, in the latter it was clearly yes. This will be discussed further later in the next chapter.

'Teamwork'

Following from the groundwork of building trust and respect, the next temporal step was to start work. Many of the social dynamics of working together were collapsed into stories about and discussions of "teamwork". As with 'getting to know you', many more nuanced issues were folded into the idea of teamwork, and both teamwork and getting to know you were more closely connected than a simple temporal progression suggests. As one researcher described the process:

> *I suspect it takes a long time to set up work teams. One, you've got to develop personal trust, that you're not going to get ripped off scientifically or they're not going to shoot you down every time you stand up. So it's about building personal relationships where they don't exist, and in many cases I think they didn't. And then you've got to build a work relationship, and in some cases your work has got to evolve to a stage where you start to put interconnections into it, not necessarily start at day one.*

This comment encapsulates the inter-relatedness of trust and respect and the technical dimensions of work. The 'getting to know you' phase and the 'working together' phase were often intimately connected. Indeed, several researchers felt that the only way to see whether they could work with someone is to actually work with them. The interconnections between the social and technical dimensions of work will be discussed extensively in the next chapter. For the remainder of this section, I will focus particularly on the social aspects of working together.

It is important to note here that trust and respect were still a large part of these social aspects. They were not checkbox characteristics that, once a person was 'ticked' as trustworthy, for example, they would stay that way. Rather, the basis of the relationships formed by trust and respect was continually reassessed through time. In the 'teamwork' phase, attention turned more toward the tools that *allowed* this continuous assessment.

There were two issues that were most prominent in discussions of assessing and maintaining relationships over time, which in turn allowed those relationships to develop. They were communication and fairness.

Communication   The importance of effective and timely communication was a common theme. This was less communication in the informal sense of people working together at a micro level, and more as a formal tool for building and maintaining relationships. All levels of leadership, from task and project leaders through to CEOs recognised the need for appropriate communication. For example:

> *I'd really like to know how much people feel they belong to the CRC, are there good internal communication networks, do they feel as though they are being kept up-to-date with current developments in the CRC through their teams, or team leaders, do they feel that they can have a say that will be recognised?*

and

> *one thing I've been trying to do, probably not as well as I should have, it may sound like a cliché, but keeping communications open. I try and get a newsletter out either way saying what's going on. Then people know what's going on, and people feel that there's some kind of interest in them, that they know what's happening up above.*

In the absence of any communication, it was common for people to think that nothing was happening. As one frustrated researcher described it:

> *I basically haven't had a clue what's been going on between last year's workshop and this year's, and there seems to be prevalent view around that 'God, nothing's happened'. We all got together and spent a week putting this stuff together and nothing has happened. Now, I'm sure a lot has happened but... even though I'm based in the same building, and my boss is here and I see him nearly every day I still have no idea what's going on.*

This perception was sometimes quite mistaken, with potentially serious consequences. For example, in the Coastal CRC, following the launch of the Centre in December 1999, there was a period of between 12 and 18 months where operational systems were designed and implemented, negotiations over resources were taking place at a high level, project proposals were being considered, the Board members were getting to know each other and the CRC, the Executive Management Group was established, key stakeholders were being sought, and so on. Much of this activity was administrative, in the sense that it involved primarily the administrative core of the CRC, and so little of this activity was communicated to researchers and other stakeholder partners who had been involved in the bid and had submitted project proposals. The consequences of this insufficient communication at a stage when relationships among CRC participants between those participants and their stakeholders were often new and tenuous, were significant. As one researcher remarked:

> there have been so many workshops in the early part, saying 'we're going to do this, we're going to do that', and that was in March, April. And [stakeholders] are still waiting for these experts to come up and do something. Now that's, you know, going on 12 months.

In other words, as communication was a key means for assessing whether the CRC was going to be trustworthy and treat their prospective input with the respect it deserved, silence was not an absence of information—it spoke volumes. People were just as ready to learn from lack of communication as they were from the presence of it.

Over-communicating could also be counterproductive. Several researchers, especially those in leadership roles, spoke of the work involved in keeping in touch with all the relevant people. Physical distance was perceived as a barrier here, in that it was harder to coordinate people to talk, leading to greater fragmentation of work time to organise or attend meetings. Although email and teleconferencing were used extensively, they were widely recognised to be no substitute for meeting in person.

> The more you communicate in the group, I think, the more you fragment your own time to do things. There might be some people who need a lot of linkage and others would probably be a lot more effective if just left alone, and there's probably different personality types that require—and different areas of research perhaps which have different networking requirements.

and

> I guess the networking also comes with a cost in that you have to put the effort and time into communicating with a wider group of people. It depends I suppose on how interested and active and cooperative the other people are. What sort of benefit can be realised out of that....

The role of inter-personal communication in integrated research was vital. This did not need to be formal communication, but informal too. Location was significant in this context, as one researcher located in an isolated research group noted:

> we miss out on a lot of the CRC. We are always getting messages saying drinks on Monday afternoon, somebody is giving a talk about something, and you think 'gee, those people obviously have a lot better chance of integrating their work than we do because of the social contact'.

Given the geographic dispersion of the CRCs, communication from the main centres of action could also reinforce a sense of isolation for those who were distant from the centres.

Fairness and transparency A second, although less common, tool for enabling people to assess their relationships were perceptions of fairness. Fairness differed from trust and respect, as it was concerned with how that trust or respect was manifested in organisational processes and procedures. People were less concerned that events did not unfold the way they had anticipated if they could acknowledge that the processes involved in that unfolding were fair.

Perhaps unsurprisingly, this often came down to budgeting and the allocation of resources. Dividing resources was a fraught process for both CRCs. Different salaries, administrative processes, charging rates—even air travel classes were cited as contributing to the budgeting 'nightmare'. In some instances, establishing ways of allocating resources that respected all participants' status and the extent that they were likely to contribute was a somewhat creative process involving long and intense negotiations:

> *to get them even to agree and even be moderately happy was my crowning success I think. [Laugh] So it's partitioning of the dollars that is the thing. If people are reasonably happy they'll get on and do their work, but that was a bit nightmarish, and we're going to face it again at the end of year 2 or 3, whenever we reorganise again. And that takes six months or so.*

Nevertheless, the ability to argue for one allocation over another required a mechanism that was widely acknowledged as fair.

Fair procedures on their own were only useful if people knew about them. As such, the concepts of communication and fairness were linked through transparency. Transparency of processes was an important component in people's judgements of the organisation at large. Even where key individuals (CEOs or Program/Theme leaders, for example) had the CRC participants' trust and respect, if the processes weren't in place for people to see what was happening and how important decisions were being made, they could easily become disenfranchised with the CRC as a whole. Once again, the silence spoke clearly:

> *I got the impression that it was sort of a bit like commissioned research, that they went behind the scenes—I'm presuming this 'cos I don't know what happened....*

Judging appropriate levels of transparency with stakeholders could be difficult. Traditional scientific methods such as papers or technical reports were of varying use as transparent records of the research, as their transparency varied according to the technical capability of the stakeholders. What might be too simplified for some may be too complex for others. For example, in the Coastal CRC the different Management Study Areas had quite different informational requirements from the CRC. In the Fitzroy MSA, for example, many of the stakeholders were landholders and community representatives without extensive technical knowledge. In Port Curtis, the stakeholders were mostly industry representatives, including environmental managers with tertiary training and substantial experience in coastal environmental issues. In South East Queensland there was a mixture of both, with the addition of a large local council Waterways unit of 22 staff.

In this diverse setting, conventional communication strategies, including 'let's-produce-another-brochure syndrome' as one communication manager referred to it, were not likely to be sufficient. Communication managers recognised that communication planning and implementation needed to be sophisticated and ongoing, rather than the ad hoc production of a brochure or a report at the end of the research. However, even the latter was often still a challenge given the inexperience, and sometimes the reluctance, of scientists to prepare materials for non-science audiences. As one manager described it:

It's almost impossible. The desire to edit and edit and edit until it is so perfect, the desire to elaborate beyond a page, the desire to get everything to such a point that it is—it can't be work in progress because [they're worried] about the peer attack I think. They don't want their peers to laugh at them, and why would they? They are shy, they think no one will be interested in it, they don't think it's groundbreaking so it never occurs to them that people might derive value from it.

However, the perception that the credibility of the CRC was intimately connected with its communication activities was given fairly widespread support. A wide range of researchers and other participants spoke of the need for the CRC to develop it's public profile and engage in public relations activities, as well as communicating more with identified stakeholders.

Back to the future

The previous sections have primarily drawn on learning from experience as a basis for current judgements. However, as Wenger (1998) has pointed out, just as our learning stretches backward through time, so too it stretches forward—we do not simply act at the present, we act into the future. In other words, what we choose to remember from the richness of our everyday experiences reflects what we expect we will need to know in the future. The discussions of integrated research reflected this simultaneous looking backward and forward. As many examples have already indicated, looking forward was often an implicit component of the significance of trust, respect (can I work with these people in the future?), communication, fairness and transparency (how can I judge whether the trust and respect I have for my colleagues today needs to be revised to guide future decisions?). In this sense, all of the discussions so far in this chapter have been concerned with the connections of the past to the future through learning.

However, as with the layering of fairness and transparency over trust and respect, so too there are issues that the study participants themselves saw as more explicitly concerned with the future. These were most commonly expressed as expectations and visions.

Expectations   Assessments of the CRCs' performances, and people's satisfaction (or dissatisfaction) with it were often expressed in terms of how their experience fell short of their expectations. These expectations themselves were widely different, and ranged from dollars they would receive to different expectations about the way work was to be carried out in partnership.

Several researchers thought that expectations had been poorly managed in the CRCs. In the Greenhouse Accounting CRC there had been expectations among some researchers supporting the CRC bid that one of the early functions of the CRC would be to do the carbon accounting for Australia, instead of the National Carbon Accounting System administered through the Australian Greenhouse Office. When that did not eventuate, there were then expectations that the CRC would still do a substantial proportion of the NCAS work under contract. When that failed to eventuate, several participants were disgruntled:

> *It just irritates the hell out of me that we've gone to all the trouble of organising this, and people go out and learn about the issues, and then go and pick up [consultancies] as... what do you call them? Bilateral relationships, not multi-lateral.*

Similarly, some participants perceived that the Coastal CRC had, in gathering support for the bid, 'talked up' what they were going to deliver, and generated high expectations. As already discussed, the subsequent delay in getting projects funded was then all the more damaging as researcher and stakeholder expectations were already raised, then nothing happened for a period of up to 12 months.

The role of expectations was not a vacuous one—expectations were real, in the sense that people planned their activities not only around what they expected was going to happen, but when. One student, for example, had planned research around data that were expected to be available at an appropriate time to complete her studies; delays in accessing that data meant that the project had to be substantially rewritten to allow completion within the required time.

So, people's expectations of the future shaped their current activities in direct and practical ways, ways that are probably rarely noticed until the expectations are proved wrong. In an integrated research setting, expectations became more tenuous, as the diversity of people involved, and the diversity of the constraints on those people, were less well known and hence less predictable than might otherwise be the case within a single research organisation or institution. This increase in uncertainty was not widely articulated in the CRCs, but it appeared to be a significant aspect of the participants' views of the future.

Visions   In contrast to looking forward to an expected future, talk of 'visions' reflected ideas of a preferred future. Sometimes these were different, but most often visions were spoken of as significant in shaping the future. To repeat the quotation that opened the first chapter, by Coastal CRC CEO, Roger Shaw:

> *And making a difference. Yeah, that's it, can we make a difference in the coastal zone? That's a goal. That's all part of bringing people together into something that can make a difference.*

Several people referred to the importance of visions as providing a sense of leadership and direction:

I think vision is the critical thing really, and that is that there aren't really a lot of people who are interested in the vision, they'd rather have someone telling them where to go. ... We've set ourselves the vision of bridging the gaps, bridging the gaps between science and policy and the only way we can do that is by having an integrated picture.

Also important was a sense of difference—visions helped to define how the CRCs were different from other research organisations, thereby creating an identity people could relate to. In the absence of a clear vision, people could easily feel as though they did not belong to anything.

In contrast, a clear vision was also a benchmark for performance. For example, part of the Coastal CRC's vision was to 'crash through on big issues' in coastal management. Some were skeptical as to whether this goal was being achieved:

are we actually doing what our mandate is and crashing through on things that require seven years of funding, the big issues that no one wants to touch, instead of defaulting back to the mundane?

So, having visions and aspirations simultaneously created expectations and benchmarks as to whether the CRCs were achieving what they were claiming.

IDENTITY IN SOCIAL BORDERLANDS

The preceding sections of this chapter have attempted to show that by highlighting the practices of doing integrated research—of integrating—the relevance of social, historical, political and cultural contexts of research can be placed within a broader framework of developing workable relationships and communities. But as the previous discussion has suggested, the borders were not only organisational, nor even inter-personal. They also affected how people understood themselves and their own activities in relation to others—in short, their sense of identity.

A useful concept here may be that of social 'borderlands'. Borderlands exist wherever any single person simultaneously inhabits two (or more) communities (Bowker and Star, 1999). The models of the previous chapter illustrated several ways people were attempting to make sense of how the CRC *as an organisation* was attempting to work across the institutional and organisational borders. At the aggregated levels of groups, programs, organisations, and so forth, it made sense to most people to understand activity across these borders as the flow of decontextualised information.

However, as most people divided their work time between the CRC and their home organisations, it was not only the CRCs, but the *individuals* who were participating across organisational borders. At the individual level, it was the personal and inter-personal aspects of working across borders that people were struggling to make sense of. At this scale, the technical issues of bringing together decontextualised knowledge did not help participants account for the social and personal challenges of working in social, scientific, and organisational borderlands.

Multiple identities as 'hat juggling'

Several people used the metaphor of 'changing hats' to describe their multiple roles both within and outside the CRCs. This was an acknowledgement of their position on the borders of different groups, and the challenges of dual, or even multiple, responsibilities. However, even when the metaphor was not used, it was possible to discern when people switched their perspective according to their different roles.

The 'hats' were also indicators of the ways people most strongly identified with their work, how they saw themselves, and what they perceived as being relevant about themselves and their work in relation to others. These categories included institution, organisation, and discipline.

Institution   As mentioned earlier, the term 'institution' is used here with respect to broader societal structures, such as science, community, government, business and industry. Institutional hats were swapped as borders were crossed—for example, stakeholders getting involved in research projects were moving across the borders that separate science and government. As one stakeholder participant described it:

> *you need to be part of a research team, with the users being involved equally. I've allocated 20% of my time to put on a researcher hat and be part of a research team.*

Similarly, members of the CRC Boards were engaged in a delicate hat-swapping, border-crossing exercise, especially where they participated on the Board as representing stakeholders, not research organisations. For example, both the Australian Greenhouse Office and the Brisbane City Council each had members on the Boards of the Greenhouse Accounting CRC and the Coastal CRC respectively. All Board members were in the somewhat ironic situation of being expected to put their home affiliations aside in the interests of the CRC, yet it was their home affiliations that justified their place on the Board. While these connections and duplications could be used positively, they were also occasional sources of suspicion, if not direct conflict.

Organisation   The most common hat-swapping was between organisations, especially the CRC and participants' home organisations:

> *I'm wearing three hats—four hats: first hat is [on the CRC management team], second hat is project and task leader, third hat is as a representative of [my organisation's] Board member, and the fourth hat is that I do university work as well.*

Again, the identification of the different responsibilities was seen both in terms of conflict, and as a positive source of cross-fertilisation. For example, one researcher spoke of the benefits of having two 'hats':

> *I think what the CRC has provided for me, I guess is that exposure to that whole range of other people outside of DNR [Department of Natural Resources] and... —I'll put my DNR hat on—and there is a lot of place paralleling between what DNR is doing and what the CRC is doing as well.*

In contrast, another saw it as a source of unease, again reflecting the tension between competition and collaboration:

> *if a third party wants you to do things [consultancies] do you put your CRC hat on and say 'I'd like to do this on behalf of the CRC', or do you use the 20 or 30 or 50 percent of your time not committed to the CRC [to] deal with this on my other side?*

So while the opportunity to swap hats may be part of the whole rationale for CRCs, the benefits were not necessarily readily realised.

Discipline Somewhat surprisingly, disciplinary affiliations featured only rarely as a relevant boundary, suggesting that it was not as strong a source of identity as home organisations were. This runs counter to a large body of literature that regards disciplinary background as a major factor in how researchers see themselves and their work. In these cases, it may be that the researchers involved were already experienced in inter-disciplinary work, and did not find it socially threatening; alternatively, perhaps their disciplinary training was so much a part of their psyche or acculturation that they did not think to bring it up in con-versation. While it was not possible to canvass this issue across the range of participants, there were several indications that the former applied, rather than the latter:

> *I've gone into almost another field. I'm used to dealing with—I'm a microbiologist, and I deal with processes and microbiology in wastewater, water treatment and water quality issues, and now I'm working in a whole river... so it complements what I do, and hopefully I complement the project a bit, I'm learning a lot as I go. So I've found that really good.*

and

> *We've got a suite of really good collaborators who do ... chemistry and toxicology and modelling and all kinds of different aspects of ecosystems, to develop a better understanding of ecosystems. And communicating them effectively. So we have got graphic artists that we work with next door. So we have evolved to get ourselves out of that idea that we are only dealing with science, like-minded scientists.*

Even when not experienced in inter-disciplinary research, in general, people appeared to view the opportunity to contribute to multi- or inter-disciplinary research positively.

> *I guess sometimes when you are working on a research project in an organisation you can pretty much be sitting in your own box with a tight focus on your particular subject area, and you have limited opportunities to get involved with other researchers, and with other topics that are slightly outside your area.... I think the CRC is a nice opportunity, an excuse to have a bit wider involvement with other people slightly outside your area.*

It is likely that there was a degree of self-selection involved here, in that the CRCs attracted researchers who were open to working across disciplinary boundaries. However it may also suggest that other borders were far more significant in integrated research. In general, people appeared to view the opportunity to contribute to multi- or inter-disciplinary research positively.

Living on the edge

Joining a CRC meant moving closer to the borders of the categories discussed in the previous section, as people needed to be able to cross those borders, or straddle them, at different times. Where there was little conflict between the work of one category and another, it was relatively easy for people to maintain their dual identities—borders were fairly permeable. However, in other cases, multiple identities generated conflict and tension that could be very difficult for individuals to deal with. Interactions across the borders were rarely as simple as removing one hat and replacing it with another.

Schizophrenia: the politics of multiple identities   Holding dual or multiple identities also meant holding dual or multiple responsibilities and obligations. These were not always compatible, hence requiring balancing and occasional trade-offs, placing some people in quite delicate political situations. For example, CSIRO representatives in these CRCs, especially those in leadership roles, had to balance CSIRO's external earnings targets (as well as other stringent CSIRO requirements) with their participation in the CRCs, which rarely generated external income. This placed some CSIRO researchers in the position of having to simultaneously compete with the CRC and collaborate within it, as described earlier in this chapter.

This situation was not necessarily restricted to individuals. One researcher noted that the Coastal CRC itself was in a position of having to balance the politics of two mandates, as the requirements of the CRC Program were quite different from the goals of the CRC itself. He noted that the CRC might need to be 'a bit schizophrenic' to meet both sets of goals.

Similarly, some researchers saw 're-badging' their work as a way of dealing with the conflicts of shifting organisational affiliations while still maintaining continuity of their research, and their own sense of identity as independent scientists:

You still have to do your own thing as best you can within the context of grander plans. That don't always pan out the way you claim they are. A lot of it is in the form of—it happens everywhere—a form of re-badging. You do one thing and label it something else. ... at one stage we say, well this is to help us with our modelling of the carbon cycle, another time we say this is to help us with our modelling of impacts. Because it's the same stuff really.

In these situations, the borders create a situation where people need to be on both sides of the borders at once, and the work involved is keeping the demands of each region apart and accounted to. The borders between the two regions are maintained as separate areas of responsibility.

Participating across borders In other situations, however, the borders were not seen as boundaries to keep groups separate; rather, the CRC was an opportunity to cross some boundaries that had formerly been too difficult to cross without good reason. The CRCs provided both the reason and the means to straddle the boundaries by getting involved in new areas.

For example, several researchers enjoyed the entrée the CRC had given them to other researchers, other stakeholders, and other aspects of management:

the CRC in itself has got a lot of strength. Because there's a real push for collaboration within the CRC you can usually ring any other CRC and say 'hey, look, what do you think of that?', you've immediately got a network.

and

So [the CRC] provides access into a broader network that I wouldn't have been able to get into in any other way. And the opportunities that the meetings have provided for informal discussion, thrashing out issues.

In these cases, the borderlands were not separators but a new region to be participated in, even enjoyed!

For others, though, the experience was not so enjoyable. Several researchers saw the CRCs as involving "forced collaborations" where they did not get to choose their research partners, a prospect some resented quite strongly. In other words, while happy to participate in the borderlands, they wanted to do so on their own terms.

Again this difference can be understood as different views of the value of the process. Those who enjoyed working across traditional boundaries tended to view the process itself as a valuable experience—even if the research itself is harder and perhaps riskier in that it may not work out the way one plans, the end product is only a part of the benefit of engaging in the process. People who viewed this negatively tended to focus solely on the quality of the research output, and viewed interdisciplinary or cross-institutional work as an unnecessary complication in the attempt to do good science. This dichotomy was not absolute, of course. Those in between were sometimes simply pragmatic about the necessity of working across boundaries, and hoped that the best science would result anyway:

underlying all this we know that most scientific developments occur serendipitously out of the alert mind keeping an eye open when other objectives are being striven for. You could say, I suppose, that it's just as likely to get a serendipitous observation which leads to the next big leap forward while doing something that meets a client's needs as it is while doing some fundamental science to meet your own curiosity, perhaps.

However, there were also different perceptions or judgements as to the payoff for taking the risk of stepping beyond one's usual boundaries. Some saw the potential benefits as very high, to be balanced against the costs of *not* working across the borders. The CEO of the Coastal CRC viewed the risks in this way, and actively urged others in the CRC to do likewise. In order to be able to 'make a difference', there was no choice *but* to participate in the borderlands.

Others, however, were more conservative—indeed, the two CRCs contrasted quite strongly in this respect. For the Greenhouse Accounting CRC, the risk assessment was quite different. Participation in the borderlands might be inevitable, but the risks of being seen to be co-opted by non-scientific interests could greatly reduce the CRC's ability to make a difference in the tense political arena it needed to feed into.

So while some borderlands needed to be straddled, maintaining separate feet on each side of the border, so to speak, others could be participated within, with both benefits and disadvantages.

CREATING, DEVELOPING AND MAINTAINING BORDERLANDS

By using the borderlands metaphor I have tried to illustrate the complex layering of boundaries upon which the CRCs are placed. By creating a new space atop these layers, the CRC is essentially both a product of those borders (the presence of the borders being the rationale for their existence) and a place where their presence may fade (in the sense that they are seen to be more permeable), as illustrated in Figure 21.

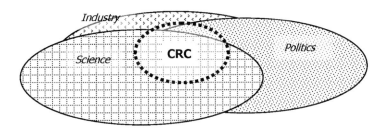

Figure 21. Overlapping borders.

But borderlands are not only created, they must be developed and maintained, and this work comes down to the people who are participating across these boundaries. The structures and models of Chapter 4 only exist insofar as the people involved create and use them to understand and articulate integrated research. As such, the differences in the historical experiences of the research participants will affect their perceptions of, and participation in, the integrated research setting. The separation of the abstract concept of integration from the contextualised practice of integrating is, despite its intuitive appeal, an artificial boundary. This boundary is embedded in historical notions of knowledge as separate from knower, science as independent of politics, and the possibility of objective knowledge.

Yet as the quotations of this chapter suggest, attempting to maintain this separation in an integrated research context has repercussions socially, politically, and personally. One way of understanding this disjuncture between the highly contextualised activity of doing integrated research from the idea of integration as the manipulation of decontextualised knowledge is as a way of categorising work activity that is *consequential*. From the perspective of science it leads to confusion, as the socio-political 'incursions' on research practice become harder to ignore, and yet cannot be granted meaning within the 'objective' scientific world. From the perspective of stakeholders it becomes a source of frustration, as scientists simply fail to understand their socio-political pressures and needs. At the individual level it can lead to identity crises and ethical dilemmas.

The consequences of categorisation can be seen in terms of the ways in which people learn from their experiences. This chapter has illustrated that people will learn rapidly from their socio-political experiences, in terms of who they will or will not trust, who they respect for their competence or values, and so forth. Chapter 4 illustrated that participants readily learned from the technical advances and developments of others to create highly sophisticated technical approaches to designing and modelling integration. In dealing with this diverse, uncertain world the technical and social views simply represent different ways of labelling *the same process* of making sense of and acting within that complexity. Yet the categorical separation of these two types of integrative activity means that there is little connection to allow learning *across* these categories. This was exemplified by one researcher who, in describing how he 'blundered into' some highly useful and productive relationships with local stakeholders, stated, "That's a kind of integration, but not an official one." Consequently, the separation of technical, 'official' integration and socio-political 'unofficial' integration persisted in the perspectives of many researchers.

It may seem odd, perhaps, that people can simultaneously offer quite differing accounts of what integration is as a concept and what integrating is as an activity, but given the hold of the Cartesian split on Western thinking this need not be so surprising. As Lave comments, "... a belief that the world is divided into contextualized and decontextualized phenomena is not merely an academic speculation that can be discarded when found theoretically inadequate or incomplete. This dualistic view of the world has a lively presence in our everyday lives" (1993, p. 6). In this way, the formalisations of Chapter 4 can also be understood as based in the personal histories of individuals (mostly) trained in the sciences, as well as the broader socio-cultural history of science as a rational, abstract, observable, puzzle-solving process. These both contribute to the 'intuitive common sense' of understanding of integration work in this way. The general acceptability of the models of Chapter 4 when they were presented to the participants in the second round of research conversations confirmed the 'comfortableness' of both the explicit manifestations of the information flow models as well as their conformation to the tacit conventions of a 'scientific' approach.

While conceptualising the problems of the gap between research and its application is widely recognised as an issue of the relationship between science and its context (for a critique of this view, see Latour, 1999), these two chapters suggest that there the gap may equally be an issue of a categorical divide between techno-scientific and socio-political activity.

However, as Latour and others have strongly argued with respect to the science-context divide (Latour, 1999; Pickering, 1995), this separation between technical and social is only a way of understanding and talking about research that makes sense to people. This distinction between social and technical aspects of activity, like that of contextualised and decontextualised phenomena referred to at the start of this chapter, is a deeply embedded, 'lively presence' in our lives. It has the effect of marginalising all other layers apart from the scientific/technical, an effect seen in the dominance of the technical information flow models of integrated research. What happens to all these 'other' layers of activity?

Bowker and Star (1999) describe the phenomenon of the missing, unclassified aspects of practice as the presence of 'residual' categories, the nameless bits of activity that fall between the gaps or overlaps across identified categories. The ubiquity of what is excluded from the categorical system, and that they remain nameless and therefore undiscussed, suggests that an analyst cannot reach an understanding of the residual nameless 'other' between only by examining the extant categories. In other words, to start to articulate what goes on in between the social and technical categories of integrated research, how they interact and shape each other in the practice of integrated research, I need to take a different perspective from that used in this and the previous chapter. In these chapters I have drawn concepts from those that the participants used, from their descriptions of their experiences. While these categories are illuminating in certain ways, to go beyond them requires an alternative approach.

Seeking a new view

The conceptual separation of the social and the technical aspects of integrated research suggests that these could be separated in practice, that the technical issues of research could be isolated from the social and political milieu with which it was (somehow) connected. However, the tensions and conflicts discussed in this chapter illustrate that such separation was, in some cases, problematic. To go further, it can be argued that the *usefulness* of separating of social and technical aspects of research—which is largely unquestioned within the conventional scientific institution—is markedly diminished in the context of integrated research that seeks to make a difference in action.

Yet ability of the participants to engage in integrated research indicates that people had developed ways of reconciling the two in practice. This raises the question, how does the category system of social and technical understandings of integrated research relate to research practice? How do researchers learn to talk with colleagues from other disciplines, policy-makers, community groups, the media, industry partners and others in ways that build upon trust and respect, exemplify transparency, are sensitive to their presence on the borders, and allow them to maintain an identity that continues to make sense—while still doing quality research? What may the technical biophysical integrative context tell researchers about the socio-political context? What may issues of identity and inter-personal relationships suggest about the context in which the research may eventually be put into action?

These questions bring us to the brink of the 'conceptual no-man's-land' discussed in the opening chapter, as well as to the Cartesian split between rational thought and the messy, confusing realm of action. Indeed, the last two chapters suggest that these are one and the same—the breach between the rationality and abstraction of integrated research models and the historical, political and social melee within which research must participate to make a difference.

Yet the same stories, re-told, also offer answers to such questions—small-scale answers, based on individual or small group experiences and learnings of how to manage working across such a divide. In the next Chapter I will turn from fragmented accounts of how people describe the benefits and conflicts of integrated research towards stories, or vignettes, of how the process of doing the research worked. These stories will be used to illustrate *connections* between the two categories of 'technical' and 'social' aspects of integration, and how they were understood, and thereby explore the residual category of all that lies outside or across these two categories. In particular, I will seek to highlight the (often implicit) ways in which these aspects of experience *informed* each other, and became sources of learning. By focusing on connections between the technical and social dimensions of integrated research (are there any connections that people used more or less consistently?) the lessons learned by individuals or small groups may offer some more general constructs, new category systems, that may start to fill in the 'conceptual no-man's-land' between science and everybody else.

CHAPTER 6

CROSSING THE GREAT DIVIDE: ACTION CONTEXTS

FROM RATIONAL MODELS TO INTUITIVE INTERACTION TO ACTIVITY

In Chapter 4, the constructs people used when discussing the concept of integrated research were found to be largely rational, technical models based mostly upon a metaphor of information flowing from one agent to another. In Chapter 5, by focusing on how people described their experiences of integrated research, a picture of research as an intuitive, political, historical and social process emerged. However, as the conflicts and confusions noted in Chapter 5 suggested, maintaining a border between the technical and social aspects of research was not a particularly useful strategy for understanding integrated research practice. Like all simple categorical systems, the social/technical divide highlights the differences between those aspects of integrated research defined by the categories, but fails to account for other significant aspects. This chapter seeks to answer two questions: what does the social/technical category structure, as a way of articulating integrated research, conceal, hide or obscure? And what are the 'nameless' aspects of integrated research practice—the skills, perspectives, concepts or tools—that people used to make sense of their interactions across the social/technical divide?

From data to description

To answer those questions, the analysis presented here is structured differently from that of the previous chapters. Whereas Chapters 4 and 5 reported on the two different types of thematic investigation across all of the research conversations, in this chapter I shift to the use of vignettes, or stories.

The stories represent a different unit of analysis to the isolated comments focused on in Chapters 4 and 5, to the analysis of whole events, or projects. This different unit of analysis was used to focus attention on the ways in which people participated in and made sense of whole processes of working together, rather than their descriptions of the concept or experiences of integrated research. The stories offer a different perspective on the ways people actually navigated the social/technical categorical divide in integrated research practice.

The stories reported in this chapter were selected to illustrate a diverse suite of integrative research contexts and approaches. They are told initially from the perspective of the social/technical divide: how the social/technical boundary became perceived to be problematic, and different responses to those problems. Each of the stories are then re-told to emphasise alternative ways people came to understand and characterise their relationships. The categories that emerge from these stories will be developed further in Chapter 7, where I will explore how the major themes that emerge from this analysis may offer useful concepts that can be used to articulate and negotiate what integrated research is and how it can or should be engaged in.

It should be noted here that in this chapter the concept of 'social/technical boundaries' is used as a simple abbreviation of the more complex divides discussed in the previous two chapters. 'Social' includes political and organisational issues, 'technical' includes issues of science and information flows.

STORY 1. THE GREENHOUSE ACCOUNTING CRC'S STRATEGIC PLAN

As mentioned in Chapter 3, the rationale behind the funding of the Greenhouse Accounting CRC lay firmly in the negotiations surrounding the establishment of the Kyoto Protocol. Consequently the CRC was born into a hotly contested political arena, an arena in which science had played—and continues to play—a key role in high-level policy decisions. This was, of course, no coincidence. It was the *political* need for scientific input that rendered this particular branch of science so relevant at the point that the CRC bids were being considered.

The original structure and research plan of the Centre had been developed for the CRC bid in 1998. In a rapidly moving political arena, by 2000 it was already considered to be out-of-date. This had been anticipated and planned for—the Centre's initial projects had been funded for two to three years, so they would be completed in 2001, ready for revision. Consequently, in 2000 a strategic planning process to develop the directions of the new research projects was started. The process was extensive, and included consultation with many researchers at all levels of the organisation, as well as input from the management team and the Board.

The new 'Program D'

One of the main outcomes of the plan was the restructuring of the four original programs (Programs 1 through 4, introduced in Chapter 3) to four adjusted programs (Programs A through D). The changes were mostly to rearrange management workloads, and to cluster the research projects to better integrate those aspects of carbon accounting that had been separated under the former structure.

However, within that larger restructure, two issues emerged in the strategic plan that were particularly significant with respect to integration. The first was a new program, Program D, 'Science Applications and Outreach'. This program's goal was: "To respond to requests from participants and stakeholders for rapid scientific responses to issues of current and future relevance" (CRC for Greenhouse Accounting, 2001b, p. 19). It was comprised of three new projects—Good Practice, Carbon Scenario Analysis for Land Management Change, and Issues in Carbon Accounting—which in total were allocated approximately 10% of the Centre's annual research budget.

The origins of this new Program were both simple and complex. The simple origin was a recognition by the CRC planning team that the Centre needed to be more flexible and responsive to the needs of their stakeholders including, but not restricted to, government. This was an area they felt they could improve, and as research results were beginning to emerge from the projects, it was timely for a formal mechanism and resources to be allocated.

The more complex origins lay in a broader recognition that the CRC could not anticipate what the needs of their stakeholder communities were going to be. The relationship between the technical work of the CRC and the social context within which its research was situated was central to the Centre's ability to justify how the research was relevant and integrated with national needs and priorities. Yet the national (and stakeholder) needs were themselves located within an international political arena that the Centre also needed to consider, as changes internationally would inevitably shape the Australian Federal Government's priorities, which would then feed through to other tiers of government and to industry. This relationship was well understood by the Centre's management, and their strategic planning process had been timed so that outcomes from the 6th Conference of Parties in The Hague, the Netherlands, in November 2000 could be incorporated into their future activities. The relationship between the technical planning and the social context became problematic when international negotiations stalled. There was an hiatus when the United States withdrew from the Kyoto Protocol, at which point the negotiations were suspended for eight months.

These negotiations were crucial to the Centre as they included a decision on whether carbon sinks were to remain in or out of the Protocol, a decision that would strongly influence policies concerned with greenhouse accounting issues. Consequently the CRC suspended its own planning process. When it was clear that the international negotiations were not going to be resumed until July 2001, it was decided that the CRC needed to plan their activities despite this uncertainty, and the development of the strategic plan continued. As the CEO reported:

> Even though the international climate change negotiations on implementation rules for the Protocol have not been completed... the new science program and management structure [in the Strategic Plan] will equip our Centre to provide high quality research outputs that will be relevant in building up Australia's greenhouse accounting capability and for informing the development of greenhouse mitigation options for Australia's land systems. (CRC for Greenhouse Accounting, 2001a, p. 5)

Program D was a legacy of this situation. The uncertainty of the political situation was such that the context within which the science could claim relevance was highly dynamic, and, to a significant extent, unpredictable. The first project, Good Practice Guidelines, was based on the relatively certain scenario that the Intergovernmental Panel on Climate Change would go ahead with an existing plan to develop good practice guidelines for carbon accounting, regardless of the political outcomes. In other respects, however, it was extremely difficult for the Centre to plan a research program that would achieve integration with policy-makers and industry, when these partners too were facing a political context that was continually shifting. Consequently the 'Issues in Carbon Accounting' project served as a space for dealing with new issues. Having a ready-to-hand 'task force' that could respond to rapid change, while still maintaining a more conventional research program was one way of coping with this uncertainty.

The collaborative fund

The second integrative string to the strategic bow sought to overcome an entirely different social/technical problem. As noted in Chapter 3, the structure of the CRC was roughly that of research programs themed by technical issues, designed to feed into an integrative program that would be relevant to policy-makers and other stakeholders (see Figure 13 p. 66) Originally, these programs and the projects within them had also served to reinforce organisational or institutional boundaries. It was noted in Chapter 5, for example, that Program 1 was almost entirely comprised of researchers from one research school of one university. Over the first few years of the Centre, these programs served their purpose of generating research results quickly. Two years in, however, as members got to know each other socially via the CRC and saw opportunities for interesting shared work, they became increasingly frustrated by this structure. Several members saw it as stifling their ability to work across boundaries, limiting their creativity and potential synergy between groups. The programs had become silos.

Consequently there was 'bottom-up' pressure from researchers, as well as 'top-down' pressure from the CRC Program to demonstrate integration across these boundaries. It was decided that the Strategic Plan needed to incorporate some mechanism to facilitate researchers working across programs, as well as institutions and organisations. After consultation with the researchers and the Board, the Centre's CEO established a 'collaborative fund'. The collaborative fund was about 10% of the CRC annual budget, a pool that members could apply to for collaborative research. Proposals were assessed and awarded primarily by the CEO.

The collaborative fund allowed researchers to form their own research groups according to their own criteria—as one member expressed it, it enabled 'organic' projects to flourish. Decisions of who to work with were made by the researchers themselves, and so could include both social and technical considerations. This was a sharp distinction from the formal, structural approach that grouped people together solely on the basis of their technical compatibility, without consideration of their social compatibility, that had dominated the early planning of the Centre. This second integrative strategy began to pave the way for inter-disciplinary, inter-organisational, and inter-programmatic research that could access and build on the social strength of the CRC.

Implications for integration

Both of these strategies created pools of resources that could be used to support interactions where socio-political contexts could be incorporated. The new

Program D specifically aimed to respond to the needs of political and other stakeholders, and thereby facilitated integration between science and other institutions. The collaborative fund was designed to encourage greater interaction among CRC members who were not in the same home organisation or program, and as such facilitated integration within the research institution.

Yet these 'integrative' strategies that intertwined the social and technical were not ends in themselves (although they no doubt served a useful purpose in enabling the Centre to 'demonstrate' integration in its reviews). Rather, they were means to achieving a different end associated with the integrative theme: improving the CRC's capacity to bring about change. Program D sought to do this more directly, by engaging with stakeholders/decision-makers, with the project participants effectively acting as brokers between them and the Centre. The collaborative fund sought to bring about change within the academic research sphere; by facilitating more 'serendipitous' collaborations among CRC members, it was hoped that innovative approaches to integrating different research under common frameworks would emerge.

These two different contexts of change, academic development and socio-political change, were not independent. Implicit in the desire to further integrated research within the Centre was the further aim to develop integrated products that would be useful to managers and policy-makers. Consequently the immediate academic aim 'fed through' to the CRC's ability to influence decisions beyond the academic realm.

In summary then, both strategies recognised that a purely technical approach to producing the best science was insufficient on its own to bring about change outside academia—it even hampered the CRC researchers' abilities to do innovative work that crossed the many boundaries within science. These two contexts of change, academic and socio-political were connected, as it was perceived that the latter would be facilitated by more integrative science.

STORY 2. THE COASTAL CRC AND THE SOUTH EAST QUEENSLAND STUDY

Coastal research in South East Queensland is a lively and complex scene. Biophysically this area consists of several rivers, that flow either into the Brisbane River, which itself flows through the city of Brisbane and into Moreton Bay, or directly into Moreton Bay. The population of the South East Queensland area is about 3 million, with high population growth in the coastal areas. Industrial pollution, eutrophication, and a range of ecological problems have been identified in the area, with some areas unsafe for recreational use. In addition, in recent years a significant population of dugong (a sea mammal similar in size and appearance to a manatee, listed as 'vulnerable' by the World Conservation Union, IUCN, 2002) took up residence in the outer Moreton Bay. In 1994 the first stages of the Moreton Bay Study, later the South East Queensland Regional Water Quality and Management Study (the SEQ Study), commenced. This was a large-scale research program aimed at generating the understanding that would ultimately allow the authorities to 'clean up' the South East Queensland waterways.

The SEQ Study was set up as a semi-independent agency, supported by an alliance of 7 local councils upstream and along the coast of South East Queensland, which contributed funding and resources for the research. The SEQ Study was a high-profile research program, which formed the basis of a prominent 'user-friendly' public awareness campaign entitled 'Healthy Waterways'. This campaign emphasised the implications of the research for how people can contribute to the health of the waterways, including a 'Crew-members guide' that gave a series of 'Report Cards' on different areas in the bay, and used active language to encourage people to participate in the implementation of the strategy (Moreton Bay Catchment Water Quality Management Strategy Team, 1998).

When the Coastal CRC started, the SEQ Study was already well into its third stage, and had established a strong public profile. It had developed its own administrative practices and the politics of the local government councils were already complex. The Study had previously used purchaser–provider arrangements, contracting research providers on the basis of competitive bid tender to meet the research priorities of the Study. In early 2000, six projects in Stage 3 of the SEQ study were preferentially granted to the new CRC, to a total value of around AU$1.4 million. This was largely the result of key research leaders of the SEQ Study also becoming heavily involved in the Coastal CRC, and lobbying for the SEQ Study to support the CRC in this way.

From the outset, the relationship between the CRC and the SEQ Study was unsteady. The purchaser–provider model was rejected by the CRC, as it did not allow the CRC to have significant input into how the problems were set or how they were to be conducted. The problem-framing was carried out primarily by the contracting body, the SEQ Study, rather than by the contractors, the CRC. The CRC preferred to work 'in partnership' so that there could be some integration between the goals and resources of the CRC with the goals of the SEQ Study. However, the Centre's efforts to develop a relationship based on mutual negotiation of project planning conflicted with the expectations and established ways of managing research of the SEQ Study.

These problems were exacerbated because the delays in granting project funding referred to in Chapter 5 impacted upon the SEQ Study heavily, as their own milestones and deadlines were threatened by the CRC's delays. Some researchers who were caught between the organisations were highly frustrated with the CRC, with one even tendering his resignation from the CRC. This was not accepted, reportedly on the basis of contractual commitments. (That researcher did resign successfully about 12 months later.) Poor communication between the organisations fuelled disenchantment on both sides.

Despite these problems, the relationship persisted. As the CRC's technical project management systems began to operate effectively, the projects (some of which had already begun without CRC financial support) settled down to work. Relationships calmed, and many of the problems were resolved. Eventually, in late 2001, the CRC appointed a new coordinator to liaise directly between the CRC and the SEQ Study. From this point, it was hoped that work would begin to build at a rapid pace.

Social/technical divide

In terms of social and technical forms of integration, and the connections between them, this story illustrates an instance where the technical integration between the CRC and the SEQ Study, the benefits of which were not questioned, almost gave way under the collapse of socio-administrative aspects of their relationship. The categories of 'purchaser' and provider' became contested ground, as the CRC felt that being a 'provider' rather than a 'partner' was inappropriate to their goals. The description closed off opportunities for the CRC to contribute to how the problems were framed, which limited the ways in which the work could be conducted and the benefits that the CRC could get from doing the research (apart from the external funding). For example, 'science for science sake' was actively excluded, as it was quite clear that the SEQ Study were not interested in supporting or participating in the production of academic papers and other conventional research products.

One of the most significant points of this story is that the relationship between the two groups persisted, despite considerable conflict and acrimony, indicating that the 'integrative' structure of the Management Study Areas served to hold the relationships together. The formality of the Centre's commitment to these Areas— that they were written into the CRC's contracts, that they had been sanctioned by the Board, and so forth—meant that the CRC could not walk away. Similarly, of course, the SEQ Study was firmly entrenched in the same area, and was a highly desirable partner to the CRC as they were a key link between the research and local government decision-makers. However, in this instance, the formalisation of the relationships meant that the shared commitment was far more structured, more forceful than 'organic'. While this sufficed to see the relationship through the crisis described in this story, it was only when the person who was dedicated to working between the organisations was appointed that 'real' social cohesion—trust, respect, etc.—could start to be rebuilt.

Implications for integration

The SEQ Study offered the CRC two important opportunities. One was the opportunity of funding some major research projects that the Centre on its own would not have been able to undertake. The second was the opportunity to bring about change in this particular area of Queensland. Consequently the collapse of social relationships between them threatened the possibility of a partnership that could be highly productive both in terms of the technical work the Study would fund, and in terms of the entrée they could grant that work into local decision-making arenas.

Essentially, while it may have been possible for the CRC to sever ties and do some of the research independently (if alternative funding sources could be found), simply doing the research would not have ensured it would be taken up by managers. The SEQ Study, by setting a research agenda that had been devised in consultation with the local councils and agencies, had created research scenarios in which the science could be integrated into policy and even directly into management. In other words, the SEQ Study had already done a lot of the leg-work in matching the socio-political management scenarios—the things policy-makers or other resource managers wanted or needed to know about their particular coastal environment—with a technical research issue, through the commissioned research projects. However, just as importantly, they had built a social network of local resource managers who had been involved in their process and were committed (perhaps loosely) to change on the basis of the research outcomes.

In summary then, for the CRC to have influence and be able to bring about change in the management of the SEQ area, they needed to build a workable relationship with the SEQ Study. While the formal agreements could 'stand in for' genuine social cohesion (including trust, respect, etc.) in holding the relationship together over the short term and thereby allow the technical work to commence, over the longer term this was not likely to be sustainable. Yet the main point of the relationship itself was a way for the CRC to gain access to existing opportunities for achieving change in the SEQ area.

STORY 3. THE FITZROY RIVER PROJECT

The Fitzroy River project was supported by the Coastal CRC, and led by a researcher based at CSIRO in Canberra, Australian Capital Territory. The research location, the Fitzroy River, is a major river system located in Central Queensland, some 2000 kilometres to the north-west of Canberra (by road). The project team was comprised of six researchers who were located throughout Australia. Locally, the research was supported by the Coastal CRC's Management Study Area infrastructure based in Rockhampton. This mostly consisted of two staff members (later expanded to three) who were primarily employed by the Queensland Department of Natural Resources, and a vehicle. Other local resources could be negotiated through the Management Study Area Coordinator.

In many respects the Fitzroy River project was quite conventional. Its goals were overwhelmingly scientific, although they did acknowledge management applications: "...to develop a quantitative understanding of the dynamics of carbon, and major nutrients in the estuary and to quantify the major inputs of nutrients and sediment to the estuary [designed to] underpin the construction of quantitative predictive models to aid in resolution of these complicated land management issues" (Ford, 2001). The research was instigated by initial negotiations between CSIRO and the Coastal CRC in the early stages of the CRC's planning.

Early in the development of the project, some limitations became apparent. In particular, to be able to account for the tidal variability of the river system, the researchers needed to have automated measuring equipment in the river over longer periods of time than the research team or Fitzroy support staff could commit. As data were in large part collected using expensive scientific equipment, and risks of vandalism and theft meant that these instruments could not be left unattended, it was initially thought that the data collection period would be restricted to when the research team was physically present, not only in the region, but on the river. Given the geographic dispersion of the research team noted earlier, team visits were infrequent, so data collection would likewise be sporadic. Also, having a research boat on the river was expensive, and other commitments of the research team meant that having a researcher constantly in the field, especially in this relatively remote location, was not possible.

Given this situation, the project leader searched for ways of collecting the data while the actual project team was away. With the aid of a network of other researchers well established in the area, the local fishing community was asked to help by keeping the scientific equipment and taking it out on their much more frequent trips along the river. By the account of the project leader, the fishers they approached were keen, enthusiastic and "utterly reliable" in this task. They were shown how to calibrate the instruments and collect the data—in effect making them pseudo-technicians in the project.

While the local people were becoming involved in the research, the researchers were simultaneously becoming involved in the local community—there was a reciprocal process of membership going on. From understanding and sharing jokes and banter to meeting people in the supermarket, the project leader was also learning how to become a member of the local community. For example, there were often jokes made about his status as a esoteric scientist: "I used to bristle at that, but now I realise they were just having a lend of me." Similarly, he was invited to participate in local cultural practices, perhaps as a 'test' to see how far the newcomer would cross the boundaries. The project leader recalled the fishers offering him 'rum on the rocks' at seven in the morning (rum drinking is a celebrated pastime in Central and Northern Queensland): "I had to refuse, but they think I'm a bit of a wimp, I dare say." These accounts highlight the negotiative, mutual engagement that characterised the process of participating in both research and local communities simultaneously.

The success of this early collaboration with the local fishing community led the researchers to try to expand their network of data collectors later in the project. At the conclusion of my study, the project leader was negotiating with other members of the fishing community, and was confident they would be able to expand their data collection significantly. The original fishers were reportedly 'proud of their involvement', and keenly interested in the outcomes of the project. In the words of the project leader "we couldn't do it without them".

The social/technical divide

This story illustrated that the right technical information flow—the data collection on the river was crucial to the group achieving the scientific goals of the research—could be achieved by good social relationships with the local community. However, this project did not use any formal means for developing their relationship across the boundaries, relying rather on pragmatic needs, local contacts and the good will of the local people involved. The reliance on good will (the fishers were not being paid, for example) highlighted the 'socialness' of this arrangement—the fishing community could easily have walked away if they felt inclined to do so.

CHAPTER 6

Implications for integration

The implications for integration of this story are interesting as this was, largely, an academic research project. Consequently, in this case the commonly assumed model that science flows through to bring about changes in local communities was turned on its head. The participation of the community led to a change in what was scientifically possible. In other words, by integrating the team's technical needs with their social participation in the community, the opportunities for the research project to achieve academic change were enhanced.

STORY 4. THE 'CORE SITES' CONTROVERSY

At their 2001 Annual Science Meeting, the Greenhouse Accounting CRC was seeking ways to facilitate more 'integrated' research outputs. At that meeting several researchers proposed one way of achieving that would be through the CRC developing 'core sites'. Core sites were physical locations where several different research projects could be carried out simultaneously. This would, it was argued, allow researchers to cross-check across research results, and to construct models that could incorporate scientific information from the microscopic to the ecosystem scale. With the research dispersed across different sites, it was difficult to estimate differences attributable to the varying locations, which increased the uncertainty in modelling.

Most people involved applauded the idea of core sites: it reduced scientific uncertainty and facilitated interdisciplinary collaboration. Importantly, it was also viewed by many as a way of building social cohesion among the CRC researchers, which, it was generally agreed, was under-developed in this Centre. Social cohesion would, in turn, facilitate more collaborative, interdisciplinary work, as people got to know each other better.

The particular site that had been suggested was open woodland in central Queensland, as there had already been some relevant research carried out there that could be incorporated or serve as a basis for further research, and because open woodlands were a significant ecosystem type in terms of carbon accounting and land management. It also served the political purpose of reducing the appearance of the CRC as being 'Canberra-centric', that is, with most of the resources concentrated around the CRC headquarters. Efficiencies could also be gained as technical staff stationed at the core site could assist several different CRC projects at once; infrastructure could be used to support or service all the research in the same place. Also CRC researchers would regularly meet in person on site rather than having to travel or settle for teleconferencing.

Several issues were raised over the course of the discussion that placed the scientific processes in a larger framework of resource allocation, relationship building, safety and capability. For example, a significant argument centred on the role of technical staff to be based at the core site. It was clear that at least two people would need to be stationed at the site on a continuous basis, for occupational health and safety reasons. This would require a significant investment in infrastructure—

accommodation would need to be provided, for example. Questions were also raised concerning the ethics of placing people permanently in isolated areas to support scientific work. Several CRC participants believed that this was not tenable on the grounds that isolation was often very difficult for people to cope with, even when two people were placed there.

The proposal was debated at some length. Although the final decision rested with the Board, their consideration was reportedly strongly swayed by the views of the researchers. The researchers were concerned with cost—while it may have been more efficient to have some infrastructure on site, those gains were mitigated by the estimated costs of getting researchers to and from the sites, the costs of housing two technical staff there, and the costs of moving established research projects to a new location. In addition, several participants regarded the hoped-for social benefits as uncertain. While a core site may offer opportunities for social interaction when all the researchers were there, it was acknowledged that researchers were unlikely to spend large amounts of time at the site, and coordinating teams so that they were there simultaneously would be difficult. So while the potential for the core site to enhance social interaction among researchers was acknowledged, it was unlikely to be fully realised.

Hence the idea was rejected.

The social/technical divide

The physical location was to provide a fundamental link between the information flows and the building of social relationships. In terms of improving the flow of information across disciplinary categories, sharing the same location was a strategy to reduce the indeterminacy between them. For example, shared physical location eliminated several types of variability across research projects, including climate (rainfall, temperature, etc.), soil types, vegetation types, management history and so on. While there was no guarantee that different researchers would concur over the interpretation of their results, at least there was a degree of physical sameness from which any discussions could proceed—the boundaries of the jigsaw puzzle pieces were more sharply defined. Hence although the boundaries between disciplines were not abolished or rendered insignificant by having research based in a common space, they would be rendered more visible, and more negotiable.

The connections between these technical considerations and the socio-political ones presented a more complex picture. Indeed, the social arguments tended to remain fairly vague in the discussion, but they appeared to be based on the notion that joint field work creates a link that between work life and social life. Under field conditions (particularly isolated field sites, such as the one proposed) the strict co-location in space and time means that people need to rely on each other for social contact. Along with their research work activities, other activities—including coordinating food and cooking, washing dishes and keeping the site clean and safe—also needed to be done. In short, prolonged field work creates alternative spaces for people to get to know each other 'as people', not solely as researchers or professional colleagues.

Implications for integration

This story, like the former one, was primarily concerned with integrated technical or academic outcomes, and so begs the question of 'why bother getting to know people?' The social aspects of the core site proposal were widely held to be central to its merit, which indicate that it was also widely believed that improved social relationships among the researchers would facilitate the desired academic change. There was some degree of melding between social integration and technical integation—lack of trust among individuals at the socio-political levels, which was significant in the early days of this Centre—was largely indistinguishable from lack of trust in their work. In this case, because of the complex prior histories of many of the participants, gaining social trust became a prerequisite for developing effective working relationships.

However, there was also a further step. The concern of the proposal was largely to reduce uncertainties across scales, so that the ultimate large-scale models were as robust as possible. While there is clearly a basis for this in terms of academic change—the more technically coherent a model is, the more immune it will be to peer criticism—there was also a political element in terms of the Centre's capacity to influence and bring about change. The wider the support for any particular model from researchers both within the CRC and outside it, the more likely it would be to be used in policy-making. In other words, the social trust built in the CRC could then reinforce the technical merits of the integrative work, by supporting claims for the trustworthiness *of the technical model* outside academia, thereby enhancing the Centre's ability to use the model to influence decisions.

STORY 5. DECISION SUPPORT

In terms of connections across scientific boundaries, the Coastal CRC as a whole presents an extremely complex picture. In terms of the relationship between science and action, though, one particular story stands out. It is their development and use of a Multi-Objective Decision Support System (MODSS) for coastal management. The MODSS has been introduced in Chapter 3 as one of the key components of the Centre's efforts to achieve integration. It is a flexible, computer-based technique that can incorporate different variables to facilitate cooperative evaluation of different management scenarios. Scientific information can be 'built in' to the system, and social, economic or other variables can be added as stakeholders see fit. The MODSS had been a central concept in the CRC since its original proposal (The theme that was later expanded to 'Decision Frameworks', to incorporate other integrative approaches, was originally formulated solely in terms of the MODSS (Anon, 1998).). The idea was that the MODSS would be able to incorporate most of the research results generated within other CRC programs, as a basis for community, government or industry decision-making. As such, it sat at the pinnacle of the research theme triangle, illustrated in Figure 9, and was one of the Centre's flagship projects.

The MODSS project in the Coastal CRC illustrates the use of technical modelling approaches to integrate information for decision-making, connecting research and action. MODSS was seen by the researchers as a tool for 'rationalising' complex decision scenarios through formalised, visible processes. In this way, it aimed to provide a technical vehicle through which scientific, social and economic information could be integrated, and thus provide 'integrated outcomes' to users.

At the 2000 Annual Workshop, this assumption was tested as the flagship was given a dry run, in the form of a 'fish-bowl' role-play presentation by the project team, for the other CRC researchers and visiting stakeholders. They aimed to demonstrate how the MODSS process could be used, both to help researchers visualise how they could provide input into it, and to show stakeholders how it could be applied in decision-making scenarios. A prior project had been adapted to illustrate that the MODSS process could highlight the different benefits and costs of alternative decisions, and that community and industry input could be incorporated in structured ways. The role-play was the final presentation before morning tea.

The response from the stakeholders was rather lukewarm. The amount of work that would need to go into engaging in the process fully was considerable, in terms of time and effort, and it was not immediately obvious which decision scenarios might warrant that. The ability of the system to incorporate and appropriately account for all the important factors was also doubted. For example, following the presentation, one of the local stakeholders who had attended commented that the risk with this kind of approach was that groups could go through such a convoluted process and "it will still give you an answer that's just garbage." In other words, people were sceptical that in formalising decision-making through the use of the MODSS would be sufficient to capture the full complexity of the situation (or overcome political interests and hidden agendas). Nevertheless, researchers who had used similar systems before recalled occasions where it had 'opened the eyes' of influential decision-makers to alternatives they would not have otherwise considered.

The social/technical divide

In effect, the MODSS was an attempt to extend the 'technical' domain of decision-making to incorporate the social, to bring them together under a single (technical) framework. By offering a structured process it certainly sought to improve the transparency of decision-making, and reduce the need for inter-personal trust and respect. If *the technical process* was trusted and respected, then individual differences became (in theory) irrelevant. However, encouraging trust in the process was, as the fish-bowl example showed, a highly social exercise in itself. In other words, the MODSS did not integrate the social processes into the system, but rather shifted the social burden of trust from trust among the co-participants to trust by the co-participants of the researchers who were guiding the process. Categorising political, social and cultural values alongside scientific and economic values is *itself* a social, political and cultural challenge, as well as a technical one.

Implications for integration

In this story, integrating social and technical issues under the same technical process sought to provide a unified, unequivocal force for change. In the ideal situation, the MODSS would suggest a clear course of action based on carefully weighted social, economic and environmental factors. While this was acknowledged within the CRC as an uncommon scenario, it did happen, and could sometimes be a highly effective way researchers could contribute to change processes. However, it was quite different from the conventional process whereby researchers would attempt to persuade decision-makers to take a more desirable course of action, or would respond to a decision-maker's defined 'information need', to one where the scientists are deliberately placed on an equal negotiative footing with others who have an interest in the outcomes of the decisions.

SYNTHESIS: WHAT DO THESE STORIES MEAN FOR INTEGRATED RESEARCH?

These stories have illustrated different occasions or events where the technical and social strands of integration, identified in Chapters 6 and 7 as being predominantly conceptually separate, came together in research practice. It was suggested that these events could yield insights into different categories people used to make sense of coming together through integrative research.

Apart from illustrating the interconnectedness of the social and technical aspects of research, each of the stories is quite different. Yet in terms of integration, the social and technical aspects came together through the various ways in which the researchers sought to bring about change. Change, of course, is a very broad concept, and one that is often relegated to the narrow sense of "relevance criteria" for grant applications or similar. However, these stories illustrate some of the diversity and complexity of different ways in which integrated research can be construed as bringing about change—making a difference—that can be synthesised into an overarching integrative concept.

The common theme drawn out of these stories with respect to integration is that of how people construed and participated in (or sought to participate in) change. That integrated research aims to bring about change virtually goes without saying—why would it be otherwise? The time and resources involved in integrated research suggests there has to be a payoff of some sort. However, as these stories have shown, the assumptions behind how research achieves change are complex and varied. They are obscured as a 'goes-without-saying' aspect of integrated research activity, and yet central to its purpose. Consequently the remainder of this chapter will seek to clarify the role of change in integrated research.

Action contexts: bringing about change

The stories illustrate that it made sense for the various partners to cooperate in light of how they understood their cooperation could achieve change. This has been described elsewhere in this book as an 'action context'. Action contexts are the different, changing milieus in which people can shape, influence and change activity—their own or that of others. The understandings of action contexts in each of the stories ranged from the micro-level of how their direct partners can bring about change, as in the case of the Fitzroy fishers story, to the global political arena that extended well beyond direct partners, as in the Greenhouse Accounting CRC's strategic planning.

Consequently, within the general notion of action contexts, there were several different pathways to change that can be identified in the stories.

On-the-ground change   Perhaps the most immediate sense of change in environmental research is on-the-ground change, changes in management or practice based on scientific research that directly affects the biophysical environment. The persistence of the relationship between the SEQ Study and the Coastal CRC was largely based in their shared understanding of the context of on-the-ground change. In Australia, many immediate coastal management decisions are the domain of local governments and their authorities, including, for example, the location of new housing developments, and the management of sewage and stormwater. The SEQ Study had direct links to many of these authorities in the district, and the authorities relied on the SEQ Study to provide them with the best scientific advice. Consequently, by integrating their goals and resources with those of the Study (via the contracted projects), the Coastal CRC could effect direct, on-the-ground change.

Similarly, the MODSS process being developed by the Coastal CRC also sought to influence decisions that would have on-the-ground impact. It was based on the view that many environmental decisions are stymied by a lack of transparent engagement with stakeholders and failure to grapple with technical issues. In dealing with both the technical aspects of complex environments, and the social aspects of decision-making about those environments, the MODSS served as a framework within which the two could be formally reconciled to generate robust decisions that would bring about change.

This action context is, in some respects, a panacea for environmental research. Many would argue that the whole point of environmental research is ultimately to influence the ways in which we act on-the-ground. However, this was not the only pathway to bring about change.

Political change   The second obvious category is political change. The Greenhouse Accounting CRC's Strategic Plan was a prime example of placing an entire research program within the context of global political negotiation and change. It was based on a view that the research needed to be relevant to the political commitments of the Federal Government, which in turn was shaped by the international Kyoto Protocol negotiations. The flexibility that was built into the plan reflected the uncertainty of that political context.

The Greenhouse Accounting CRC had clearly positioned itself, through its earliest mandate, within the national and international political context. They aimed to influence high-level political decisions, both within Australia and, through their senior researchers' extant connections with the Intergovernmental Panel on Climate Change scientific panels, internationally. This strategy had the potential to yield change on a large scale, but was perhaps less certain than the smaller-scale, on-the-ground change favoured by the Coastal CRC.

Academic change A third action context was academic change. Despite the emphasis on the applicability of research outcomes, both CRCs were engaged to some extent in relatively 'pure' academic projects. In the Greenhouse Accounting CRC, the core sites proposal was one of these. It aimed to combine the social and technical aspects of engaging in research, largely to generate stronger academic research outcomes.

Similarly, the Fitzroy River project was predominantly academic. There were no immediate implications for the fishers who were involved; rather, the project sought to fill a significant gap in the scientific understanding of how such large tropical river systems worked. This project was remarkable for the researchers' ability to maintain the interest and commitment of the fishing community to their project when there was no obvious gain for those volunteers.

The academic research was not, however, isolated from political change, nor was political change isolated from change on-the-ground. These stories also offer some insight into how these categories of action context may interact.

'Relevant' research and action contexts

These categories of action contexts are each grounded in some notion of 'relevant' research; what it is, how to do it, and how it is most likely to bring about change. The idea of relevance currently pervades science, especially with respect to justifying requests for funding, but as yet the concept has received little serious attention (Davenport, Leitch et al., 2003) Assumptions about relevance are deeply embedded in integrated environmental research, and the tangles these assumptions can form are rarely articulated explicitly by research participants. The idea of different categories of action contexts offers a basis for a more detailed and systematic understanding of relevance, each with different implications for funding and the ways in which research can lead to change.

Yet these categories of contexts are not isolated units—as the stories have shown, they also interact. Understanding and articulating these interactions further untangle the assumptions of relevance, and can give greater depth to negotiations of the relevance of integrated research.

These are illustrated in Figure 22.

The direct effects between attempts to do relevant research and the three identified action contexts are indicated in Figure 22 by the heavier arrows, with the potential for funding noted by the dotted arrows returning resources to the researchers.

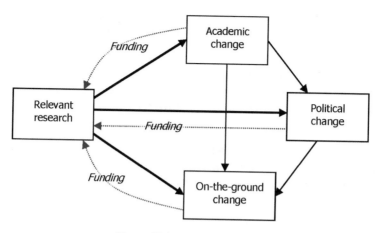

Figure 22. Pathways to change.

As noted in the previous section, change on-the-ground can be regarded, in a general sense, as the ultimate aim of environmental research. However, connecting research with specific on-the-ground action contexts often limits the scale at which the action takes place—while there are some managers responsible for large tracts of land or water resources, or large single point source pollution impacts, in many instances environmental degradation is the result of the dispersed actions of many individuals. These more dispersed actions can be altered by regulatory change or other incentives, such as economic sanctions or bonuses, which are the domain of various levels and types of governance. Hence there is a link between the political action context and change on-the-ground. In these stories, for example, the ability of the Greenhouse Accounting CRC's researchers to effect change largely lies in the translation of shifting political thought into effective policy that will change the ways in which citizens and land managers act. This is illustrated with the lighter arrow between political change and on-the-ground change in Figure 22.

Similarly, academic change is linked to other action contexts, particularly political change. It is widely held among the research community that research is more likely to have influence in political circles if it can be shown to be credible, either by virtue of the reputations of the researchers involved or by being robust to challenge. Consequently, the Greenhouse Accounting CRC's decisions to invest heavily in a fairly conventional research program, supplemented by collaborative and outreach activities, can be understood with respect to their capacity to bring about political change. The senior researchers and managers in the CRC were well aware of the controversial nature of greenhouse politics, and that 'radical' approaches to integration would be unlikely to yield action as they would be more politically contestable than conventional approaches. The efforts to reduce uncertainty in their models, exemplified by the core sites proposal, was not merely an 'academic' exercise, but a political one. This connection is also indicated by a light arrow in Figure 22.

Analysing research practice according to the different action contexts that were targeted by researchers serves as an overarching way of contextualising the diversity of approaches to integrated research that crosses the social/technical divide. The connections between research and the ways it can bring about change are part of the underlying suite of assumptions that frame the practice of integrated research. Analysing them through a structure such as that illustrated in Figure 22 can offer some insights into how and why different interactive approaches may or may not achieve the change they aspire to. Yet it also raises another swathe of questions. How can research planners more deliberately bring joint consideration of technical and social contexts into their development of integrated research projects or programs? How can different action contexts be targeted? How can integrative processes be designed? These questions form the basis of Chapter 7.

A MOVING TARGET

Trajectories of Change

FROM ACTION CONTEXTS TO INFRASTRUCTURES AND TRAJECTORIES

The stories in Chapter 6 illustrate that the practice of integrated research can be understood as processes of change with respect to different action contexts. The relationships between research and different types of action contexts were proposed as one way of beginning to unpack the tightly tangled knot of 'relevant research'. However, as the stories also showed, action contexts are not static states of affairs populated by people waiting for research to guide their decisions. They are also dynamic, changing and shifting over time. Likewise, of course, research is not a predictable linear process, but a journey of surprises and adaptation. So, while it would be a relatively simple process to assess the current state of affairs in any given action context and work towards feeding into or influencing that, by the time the appropriate research is conducted the context may have changed beyond recognition. This was clearly the scenario facing the Greenhouse Accounting CRC, for example.

Consequently this Chapter will focus on the dynamics of ongoing relationships between research and action. In it I will draw on some theoretical concepts that can help expand the notion of action contexts changing over time, and how this could be developed into a platform for negotiating, articulating, and planning integrated research. The two concepts I will use are infrastructures and trajectories. I will first discuss these in relation to conventional science, specifically, and then expand them to draw out their relevance to integrated research.

Science as information infrastructure

Science as an institution can be described as an information infrastructure. An infrastructure in this sense is a high-level structure of established philosophies, rules, techniques, criteria and more that have over time developed into an order, a suite of routines, rules and expectations that at once help to define what science is and fade invisibly into the background. Bowker and Star (1999) describe infrastructure as: being necessarily embedded into the activity of doing science, invisibly supporting research across a range of tasks beyond a single event or practice; learned as a part of becoming a member (scientist); both shapes and is shaped by conventions of practice; embodying the standards by which science is judged; having an inertia that resists change and perpetuates the strengths and weaknesses of the structure through time (p. 35). However, change is possible through internal incremental negotiation and adjustment. In other words, information infrastructures are interconnected communicating structures that allow participants to classify, assess, and relate activity to a larger conceptual and practical schema. An information infrastructure does not dictate actual processes at the level of activity, but rather can be understood as a matrix within which individuals or groups or organisations practice to be counted as doing science.

Research as trajectories

Like their biophysical counterparts (consider, for example, urban infrastructures) information infrastructures are not static, but are continually changing over time. Bowker and Star (1999) suggest that this process of change can be usefully conceptualised using the metaphor of 'trajectories'. With respect to research, then, the unfolding of activity over time forms a 'research trajectory'. Trajectories are not the only way that the temporal dimensions of research have been conceptualised, nor is the way the concept is used here the only possible way of using it. Two significant literatures will be briefly discussed before outlining the ways in which the idea of research as trajectories is articulated here.

In one of the few major sociology of science works to place time at centre stage, Pickering (1995) has discussed the temporal dimension of research extensively. His main concern is the ongoing relationships between researchers and the material world they study, in particular, how research and the material world come together in the creation and use of scientific instruments and machines. In his view, research practice is a "temporally emergent" combination of human and material agency. In other words, and to oversimplify, science is brought about by the combination of researchers' activity and the subjects they study, and the outcomes are not predetermined by either, but are a product of the ongoing interaction. While he does not use the metaphor of 'trajectory', the sense of temporal emergence indicates a very similar concept. He describes this dynamic interaction as an ongoing process, where "The practical, goal-oriented and goal-revising dialectic of resistance and accommodation is… a general feature of scientific practice" (pp. 22–23).

As noted at the start of this book, this study was primarily concerned with the relationships between people, not between people and (to use Pickering's description) the material world. This does not imply that such relationships are regarded as insignificant or misguided (as the more extreme sociologists of scientific knowledge are wont to claim, see Chapter 2). Rather, I have taken the more moderate stance that research is inevitably a product of interactions with both the material world and the broader socio-political world. It is the latter that have been the focus of this study, and in this sense that the issues that have emerged here complement, rather than contradict, Pickering's work.

The second body of literature is one in which the idea of research as a trajectory does already exist, but where its use and implications are different from the concept I will be drawing on in this chapter. This field is evolutionary economics, or 'new institutional' economics. In this literature, research trajectories are defined as paths of technological development that are shaped by significant past discoveries, but are not determined by them. They are constrained by trade-offs between economic and technological factors (Dosi, 1988). Aeronautical research, for example, has operated along two trajectories based on propeller and jet propulsion technologies. The available paths of innovation and technological development are dependent upon what has gone before, but are also dynamic and open to new breakthroughs (see, for example, Hall, 1994).

As a subset of economics, the key concern of this literature tends to be the dynamics of the relationships between innovation—in the form of technological problem-solving—and economic performance (although the concept of techno-logical trajectories has also been discussed with respect to their role in sustainable human development, see Ausubel and Langford, 1997). The concept of a research trajectory is suitably macro-economic, or at least sectoral or industry-wide in scale. In some respects, then, this can be regarded as a macro-economic version of Pickering's micro-sociological concern, that of the relationship between human systems and technology-based systems.

Both of these literatures emphasise the key features of the idea of trajectory: constrained, but not pre-determined, movement through time. While these offer different perspectives on the temporal unfolding of research in practice, the first with respect to the micro-level human–material interactions in relation to machines, and the second at the macro-level of innovation and economic growth, neither are sufficient for articulating the *social* processes that cross the no-man's land between technical and social dimensions of research. So, while they are largely complementary to this study, these conceptions of trajectory and time are not used here. Rather, this study draws from a third literature that does focus on human social interactions.

TRAJECTORIES

In this study the term trajectory is being used in much the same way as Wenger (1998) and Strauss and Corbin (cited in Bowker and Star, 1999) use it, as bio-graphical. Biographical trajectories are individual, in the sense that they are concerned with how people make sense of their activities and practices, but also inevitably social as communities provide models for how individuals can negotiate their own trajectories. As Wenger (1998) describes it: "The past, the present and the future are not in a simple straight line, but embodied in interlocked trajectories. It is a social form of temporality, where the past and the future interact as the history of a community unfolds across generations." (p. 158). For example, a person can simultaneously be a parent, a worker, a colleague, a sportsperson, and so on. These can be more or less separated (worker and colleague are closely intertwined; worker and sportsperson less so, unless in a sports team of work colleagues!), more or less punctured as life events impact on one or more trajectory (as when retrenched from work). Each of these intertwines individual action and social settings within which that action takes meaning.

In the context of integrated research, this biographical trajectory is a way of recasting the hat-juggling concept; researchers are simultaneously scientists, members of a local community, politicians and advocates, and can emphasise any one of these trajectories according to the circumstances at hand. More broadly, it offers a way of analysing and articulating the work that is required to reconcile these different demands that lie at the heart of the tensions between the concepts of integrated research as technical or social.

However, when combined with the concept of infrastructures, the concept of trajectories also offers a way of approaching the dynamics between institutions such as science and the individuals, groups, and organisations. These institutions have their own trajectories that are (to keep the picture simple for the moment) contained within these larger, infrastructural trajectories, for example, a university must conform to the accepted and recognised rules of 'good research'. This is illustrated in Figure 23.

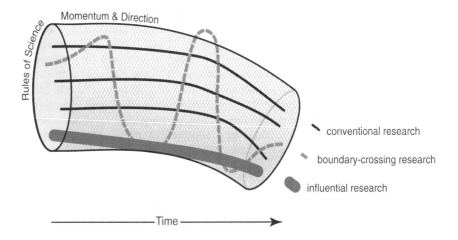

Figure 23. Research trajectories.

In Figure 23 the solid thin lines indicate individual trajectories that conform to the rules of the science matrix or boundary (conventional research), and hence reinforce the direction and momentum of the scientific infrastructure as a whole. In contrast, the thick solid line indicates that some particularly strong trajectories within the infrastructure may exert pressure for change. This can potentially destabilise the infrastructure and, over time, generate a shift in direction. The emergence of the logical positivist philosophers in the 1920s and 30s may be one such trajectory; the development of post-modern thought may prove to be another.

This concept of a larger infrastructural trajectory made up of smaller ones offers a useful way of visualising different approaches to integrated research. Integration within the sciences weaves together several previously separate trajectories, and can potentially create a strong new single strand within the science matrix, or unravel again, or exert pressure within or outside the scientific institution.

The dotted line in Figure 23 is perhaps more like integrated research that seeks to actively work with 'non-scientific' participants, working at the boundaries of the matrix. In other words, the rules of science as an infrastructural matrix are variously permeable: people operating at the edges may find opportunities to work through the gaps in the matrix, escaping from the neat structures into different spaces. To push the visual metaphor, integrated research becomes less like an electrical cable in which strands of copper are neatly twisted together and all moving in the same direction, surrounded by an impermeable plastic coating to protect it, and more like the wool of a mohair jumper, persistently escaping and going every which way despite the best attempts to spin it into a neat, strong strand.

These fibres do not, however, escape into an infrastructural vacuum, but rather pass into new infrastructures, such as government. Indeed, the different action contexts that were discussed in the previous chapter can be understood as being comprised of many such information infrastructures. It is in this way that the infrastructures concept links with the concept of boundaries; each trajectory that escapes from science forms a tenuous bridge across the boundaries created by the categories we use to label the infrastructures. Integrated research, then, can be seen as bringing the scientific infrastructure into contact with other large-scale information infrastructures, exploiting the gaps between the respective matrices to maintain their own trajectory that is only partially dependent on that of science.

Hence a shift to integrated research for the researcher is a shift away from the certainty of an established and known infrastructure to the uncertainty and opportunity afforded by leaving parts of the science infrastructure behind. The key point of trajectories is that the judgements that are exercised in hat-juggling to manage and maintain relationships within and outside the science matrix do not occur in a temporal vacuum, nor solely in an historical context. They are based on understandings of the past but *made* with respect to the anticipated *future* of the trajectories. While much of this future is unpredictable, given the indeterminacy of the trajectories noted at the start of this section, their momentum offers some degree of predictability. Without this the outcomes of all decisions would be random, with decision-making reduced not even to guesses but to a lottery.

Traditionally, within the matrix of the science information infrastructure, predictability is enhanced as the rules are known and shared. While these rules are followed and the infrastructure itself is not violated, the matrix constrains the possible futures of any research trajectory. Research that is published in respected journals is more likely to be supported by further funding; research that is replicated is more likely to withstand criticism from peers and become an accepted basis of future work, a part of the trajectory. However, research that attempts to integrate across infrastructures widens the possibility of future trajectories considerably. The research trajectories transgress the boundaries of the science matrix, increasing opportunities for playing a role in the matrices within which decisions and actions are taken, but reducing their conformity to the infrastructure that defines science.

However, as highlighted in Chapter 6, these transgressions did not result in a scientific free-for-all, but invoked the constraints of other matrices, particularly the spatial and temporal constraints of action contexts.

Trajectories of action contexts

Science is, of course, only one type of trajectory, with its own internal action context. Integrating research into other action contexts can then be represented as the collision of different information infrastructures. These collisions can be smooth and gentle or, under different circumstances, can be rough and violent. Either way, two infrastructures interact in the moments of collision, and they can twist and manipulate each other according to their own momentum. Science can be characterised as having a slow, strong momentum—it resists change and relies on the perpetuation of long-term research to contribute to the ongoing 'unfolding of science'. Even major breakthroughs or Kuhnian 'paradigm shifts', such as natural selection or quantum mechanics, take time to be absorbed and redirect the momentum of the scientific institution. In contrast, other institutions are highly punctuated, with rapid momentum and sudden shifts. The Australian Federal Government, for example, with a three-year election cycle, exemplifies what might be called middle-range rapid change. The strategies and goals that guide the trajectories of government bodies tend to be applicable over a two- to three-year time horizon, within a larger infrastructure bounded by, for example, the Constitution. Yet within that three-year period, major policy changes can still occur, such as the creation of the Australian Greenhouse Office part-way through the Federal Government's first term, in response to the Kyoto Protocol. Policy positions can change literally overnight, and the bureaucratic infrastructure that supports government, although often portrayed as lumbering and slow, appears in practice to be well used to such rapid change.

As slower trajectories collide with faster ones (slower or faster in the sense of ability to change, not passage through time!), the momentum of each exerts a kind of power to redirect the other. Bowker and Star refer to this as *torque*, "...a twisting of time lines that pull against each other, and bend or twist ... When all are aligned, there is no sense of torque or stress; when they pull against each other over a long period of time, a nightmare texture emerges" (1999, p. 27). These concepts of torque and twisting offer an interesting framework within which to consider the development of integrated research over time. Where there is no major torque, the multiple trajectories of integrated research are largely aligned, or exert only weak influence against each other. Research complements action, and vice versa. This may be the result of well-crafted research projects (or serendipitously fortunate ones) that simply fit the action context particularly well. Contracted research that is both academically rewarding and has a ready-made avenue to inform policy is one example. There are, one might imagine, ready-made holes in the two infrastructures where the demand for a particular type of input is articulated, the research can be negotiated, and the outcomes delivered. Because the work is academically valid, the incursion into the other infrastructure—bending the rules of science—are minor, and only represented by a loss of independence and the demand to produce "a result".

Alternatively, it may be that they exert little influence over each other, but happily co-exist, with no cost either. The Fitzroy River project with its cooperation between the researcher and the fishers demonstrates this process—it was not clear that employing community-based technicians to carry out the data collection represented any breach of the rules of science, and if it was then it was only a minor breach. Likewise the data collection activity for the fishers was only a minor change in their own routines, and did not impact on their major trajectory of earning a living through fishing and being part of that community. As such the joint work neatly ducked through the interstices of the aligned scientific and community trajectories, tugging a little through challenges such as the jokes about the esoteric scientist and straight rum at seven in the morning, and the fisher's capacity to learn to operate highly technical scientific instruments. As these little rough patches were worked through, the trajectory became even smoother as each settled into a comfortable routine into which all the aspects of the joint research activity—trust, friendship, reliability, work, communication—were enfolded into a shared practice that extended both backwards and forwards through time.

Of course, other encounters are not so smoothly aligned. In other instances participants in each infrastructure get caught in the twisting, as each infrastructure struggles to have its rules apply. Under these circumstances, parts of the scientific infrastructure are discarded, but not all. Parts of the political process are discarded, but not all. A new hybrid trajectory emerges, which is temporary and ad hoc, and is simultaneously beholden to the rules that form the matrix of each infrastructure *and* able to flex and even ignore those rules as it moves between the two matrices. This is illustrated in Figure 24.

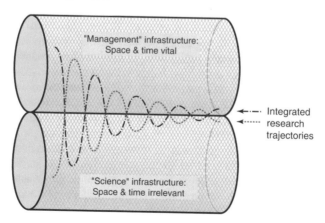

Figure 24. Hybrid research/management trajectories.

Figure 24 illustrates that what are large differences between matrices at first may 'settle' into relatively minor differences as participants become acquainted with the rules of the 'foreign' infrastructure, and establish effective ways of working together that reduces the need to flex or break sets of rules. But this is, of course, only one of many possibilities—differences could also spiral out of control, for example. If integrated research suggests that the established rules of science can be flexed and perhaps even broken, the guiding force of a single infrastructure is replaced by a range of uncertainties. Under these circumstances, how can decisions be made?

SURVIVAL SKILLS: CREATING NEW CERTAINTIES

When researchers attempt to weave the infrastructure of science with other, less familiar infrastructures, new rules need to be learned or created to restore some certainty into a much less certain situation. Researchers in the stories of Chapter 6 used several strategies to do this.

The need for speed: timetables and the SEQ study

As mentioned briefly earlier, different trajectories 'run' at different speeds, with different momentum. An extension of the trajectories construct, Bowker and Star, drawing on the work of Julius Roth (1963) use the analogy of timetables. Timetables refer to the schedules that delineate how quickly or slowly trajectories expect to meet their own goals, how far into the future those goals are set. Timetables are not, it should be stressed, a measure of productivity in which those on a faster timetable are necessarily achieving more or working more efficiently. Different timetables suit different conditions, just as it would hardly be more productive or efficient to run a peak hour bus timetable on a Sunday afternoon.

It is to be expected, then, that different research trajectories run on different timetables. An academic researcher, for example, is generally considered to have a high rate of output if she or he produces perhaps four or five published papers a year. In contrast, a parliamentary policy adviser may need to produce four or five research reports in a week or even less. A government agency research team may be 'commandeered' to focus on a particular question of current political import on a two-week turnaround, and a Senate Advisory Committee charged with investigating complex issues may take a year or more to report on a topic.

These are not, of course, accidents of institutional tardiness or freneticism. Different tasks require different depths of investigation: a rapid assessment by a policy adviser to back up a ministerial response in Parliamentary question time is a very different scenario to the in-depth analysis required of the Senate Committee. In terms of this temporal framework, different trajectories run at different frequencies, according to different timetables. However, different tasks are also constrained by the perceptions of the time necessary, or available, for them. For example, the rule that a doctorate should take three years reflects a dual perception that three years is both necessary (to do the job) and sufficient (to do it adequately).

Consequently, integrated research not only brings together different infrastructural matrices like science and government, academia and management, but also brings together different paces of activity within those matrices. This can have significant impacts on how two trajectories align.

This impact was illustrated by the relationship between the Brisbane City Council, the SEQ Study, and the Coastal CRC, as discussed in the previous chapter. The conflict between the three organisations was in large part one of unsynchronised timetables. Delays in approving the research funding through the CRC were the result of a participatory process that ensured all partners had equal say in the development and signing-off of the projects. While this was a reflection of the CRC's commitment to engage partners fully, and was in most respects an astute and effective way of building partnerships between those research partners, it was also an inevitably convoluted process that simply took a long time. While this may have been appropriate for other partners who were similarly easing into new research programs, the Council and SEQ Study were already running according to a very fast-paced timetable. Both academics and policy-makers were pushing a tightly-defined, rapid research process. Importantly, their own criteria for success were bound up in these timetables. Good research was research done "on time and on budget"—all other considerations were secondary.

The collision of the CRC and the Brisbane City Council and SEQ Study was a planned one; the SEQ Study was used as a sort of model in the early days of planning the CRC, in particular with respect to the development of the other Management Study Areas. As such they were heavily involved in the bid process and onwards. Consequently, significant attention had been paid as to how the research capability of the CRC could meet the needs of the Council and the SEQ Study. Proponents in all organisations had worked hard, and argued hard, to build the links that connected the Council and the SEQ Study to the CRC. However, within that planning, it appeared that questions of timetables were possibly underestimated. The result was significant conflict, in which trajectories would easily have been severed (such as the resignation that was not accepted) and social damage was wrought through loss of trust and respect.

While it is easy to guess according to stereotypes that the academic partners were holding up the process (a remark that was often heard in several different contexts), in this case that argument cannot be sustained. There were several very active academic staff who were ready and willing to start their work immediately; they had accepted the agreed timetable and had adapted their own trajectories to meet it. In one case temporary teaching staff had been brought in to free their time; in others doctoral students had been recruited.

Students were particularly vulnerable to timetabling conflicts, and the delays reportedly affected several early CRC students significantly. A research program of two or three years (masters and doctoral candidates respectively) does not allow a great amount of timetable flexibility as far as the research activity is concerned. One masters student, mentioned earlier, at six months into a two-year enrolment, had to rewrite her proposed project as the timetables for getting funding and access to data, which had been assured when she began, had not yet eventuated. Others had had similar experiences. Hence the easy assumption that academic timetables will drag on the other, more fast-paced research organisations is not necessarily warranted.

In the end, several researchers' commitments to the timetables of the SEQ Study in particular meant that they felt forced to advance their own timetables independently by starting the research without any direct funding from the CRC, assuming that the CRC would catch up. Once again, this was a relatively violent wrench in the trajectories that had previously worked hard to align their interests and activities, and separated the researchers from the CRC trajectory to a degree that was not easily restored. As one researcher said "I'm terminally fried on CRCs.... I'll never go down that path again." This process is illustrated in Figure 25.

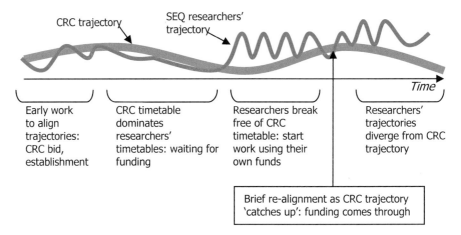

Figure 25. Timetables and trajectories: the Coastal CRC.

In Figure 25 the thick line illustrates the slow, steady timetable of the Coastal CRC, in contrast to the thin, more variable line that represents the SEQ Study. While these two went through initial periods of alignment, where similar interests and overlaps were discussed and integrated into the CRCs bid, and the SEQ Study then waited for the agreed funding to come through, the delay on this funding led to several researchers breaking away from the CRC's timetable. They started to do their research without CRC funding, which represented a significant break from the CRC's trajectory. Their activities briefly realigned when funding did come through, but the damage to the relationships were such that the SEQ researchers reduced their reliance on the Coastal CRC, and gradually diverged from them over time.

While this analysis can explain the conflict that was apparent between these partners in the early days of the CRC, it should not be suggested that the CRC's trajectory was 'too slow' in any absolute sense, nor that the researchers' timetables were 'too fast'. Indeed, even the researchers affected in this story realised that the

CRC was working hard to establish the systems that would allow the research to go ahead and that these things take time. It was the disparity in the *relative* sense between these two timetables that led to the wrenching apart of trajectories, and the difficulty, perhaps, of slowing down and speeding up. In other words, timetables are not independent of the matrices that support them. While the CRC was still building its own small-scale information infrastructure (administrative systems, budgeting processes, communication systems, and so on), its *research* timetable was inevitably slow because the infrastructure could not yet support research achievement. Its administrative timetable, it can be argued, was running very fast, but this infrastructural work was not, for the most part, directly relevant to the researchers.

The researchers, however, already had an infrastructure that could support their fast pace of research, provided the key link between the two—the funding—was made. It was this lack of synchronisation that forced the two trajectories apart.

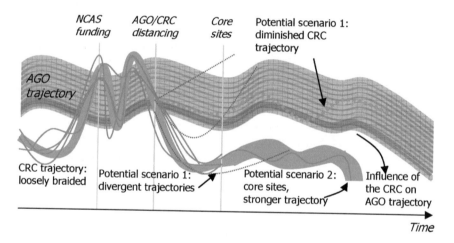

Figure 26. Trajectories under different potential scenarios.

Trajectories under different potential scenarios: the core sites proposal discussed in the previous chapter can also be understood using the concepts of trajectories, alignments and timetables. The core sites controversy illustrates a particular strategy with respect to trajectories that extends back beyond the actual core sites proposal to the relationship between the CRC and the Australian Greenhouse Office. It was widely understood in the carbon accounting research community that part of the reason science had less influence in politics than many researchers and some policy-makers thought was necessary (according to their own views of what constitutes 'good' decision-making) was due to the fragmentation of that research community. Fragmentation of research can be pictured according to the framework here as a series of trajectories that were only loosely braided, if, indeed, they were in contact at all. This is illustrated in the left-hand side of Figure 26.

With the success of the bid for the CRC, these trajectories became more closely tied together, through contact, and in some cases, new joint work. This new CRC structure pulled at most people's previous research trajectories (to varying degrees), and brought the partner agencies and organisations into closer alignment as common goals were sought. The momentum of the Australian Greenhouse Office at the time of the formation of the CRC exerted a strong torque, or influence, on the direction of the research trajectory. The strength of the pull of the AGO was enormous, given the large research budget it had at its disposal, and the CRC could potentially have tied its own trajectory very closely to the AGO to tap into this.

Many CRC members expected that this would be the case. However, others in the CRC saw that their own trajectories could be seriously compromised by the CRC joining the path of the AGO. The confusion and conflict in the early days between the CRC's potential role as a service provider to the AGO and being a more independent body reflected this, creating something akin to Bowker and Star's 'nightmare texture'. It was at this point that large parts of the CRC could have snapped their own trajectories back to their home organisations, severing ties with the CRC completely. In Figure 26 this is illustrated as the dotted lines of Potential Scenario 1, where smaller research trajectories separate from the CRC trajectory, diminishing the CRC itself, which then becomes encompassed by the AGO. The strength of the AGO's trajectory, given a large amount of torque via its rapid timetable and significant budget, led to a potentially violent encounter with the research trajectories as yet tenuously entwined in the CRC.

That this did not happen can be largely attributed to the joint decision that the AGO would regard the CRC as an independent, long-term, strategic research investment, not primarily as a short-term research service provider, although it was encouraged to take that role if it wished. This decision, announced at the 2000 Annual Science Meeting, served to loosen the bond between the AGO and the CRC significantly, and had immediate effect on the research programs. This is indicated in Figure 26 as AGO/CRC distancing.

However, the price of this gap between the CRC and the AGO was a weakening of the CRC's ability to influence government decisions. Without direct involvement through being providers for the National Carbon Accounting System, the researchers had to rely more on persuasive, innovative, ground-breaking science to capture the decision-makers' attention and to pull *their* trajectories closer to the science of the CRC. In other words, they had to build their own trajectorial strength and momentum. While they remained tenuously connected, with research trajectories that were broadly headed in the same direction, but not necessarily pulling together, this was unlikely. In other words, the administrative structure of the CRC alone was not sufficient to weave the trajectories together tightly enough to gain the *scientific* integration that was necessary for the science trajectory to pull strongly against the political one. By banding the research together more tightly, the torque that the research could exert against the politics was believed to be far greater. Hence the idea of the core sites was mooted. This is illustrated as the solid, widening bands of Potential Scenario 2 in Figure 26.

As described in Chapter 6, the core sites idea aimed to serve several purposes: scientific, logistical, technical, social and political. In effect, these were all part of the same process of building a research trajectory that was strong and cohesive enough to pull at the decision-makers, within the resource limitations of the CRC. The scientific linkages would be better, the modelling more robust, and the people involved would, through their social contact, be aware of how they could work together to achieve their shared goal of making a difference in political decision-making. In other words, in this case the scientific integration was not a matter of working across the boundaries of institutions themselves as all the people involved were researchers (although they were working across organisational boundaries); rather, they were trying to *shape the entire trajectory* of a non-scientific sector. This is illustrated as Potential Scenario 2 in Figure 26.

Figure 26 shows how trajectories can be used to structure the interpretation of these events through time, both in the past and into imagined future scenarios. It cannot make any assessment of the likelihood of the success of Potential Scenario 2—that is a matter of speculation. However, it can point to likely differences between that scenario and others. The alternative scenario that was actually implemented, that of funding collaborative projects, is likely to yield smaller, perhaps less closely-knit trajectories than would the core sites. These may then be able to influence more targeted, smaller levels of decision-making, but are less likely to exert strong torque that can lead to major change in non-research sectors.

These scenarios can also be used to offer an explanation of why the core sites did not gain widespread approval. The core sites concept tied researchers into another tightly defined trajectory, one in which each individual could only have a partial say. While this was an improvement on the AGO's trajectory, in which most researchers felt they had no say, this was still perhaps more tightly intertwined than the other trajectories (home organisations, individuals) could tolerate.

Shifting trajectory alignments

A similar strategy for coping with the vagaries of trajectories was that of changing alignments. As people left their usual infrastructural trajectories to explore new ones, as was the case in the previous section, sometimes those encountered were, for any number of reasons, in conflict with the individual research participant's own personal trajectories or the larger trajectory of the group or organisation. In the case of the Greenhouse Accounting CRC's strategic planning exercise, such a shift was identifiable in terms of the extent to which the CRC was tied into the trajectory of international negotiations. While it was well understood that the international trajectory was moving at a rapid pace initially and that the CRC would need to be versatile to keep up, it was not only the pace of the timetable that proved a challenge but the *erratic* pace. The CRC was prepared to accommodate changes to the Kyoto Protocol, especially the crucial and controversial Articles concerning carbon sinks and sequestration, in their new strategic plan—they had deliberately started with projects of two- to three-year duration so they would have the flexibility to respond to changing international circumstances. However, they were not prepared for the negotiations to stall for eight months. This is a significant period of time in a seven-year lifespan, and as the early projects were coming to the end of their first cycle, the CRC shifted its alignment away from the highly uncertain international scene to the more certain national political trajectory. Although this too was waiting for the outcomes of the negotiations, it nevertheless had a momentum that the international scene did not, a series of commitments that the CRC could coordinate with its own goals. In this way the CRC could still meet its goal of national significance, without itself stalling and losing its momentum.

Emphasis on people and personal relationships

Another strategy for coping with the uncertainty of working across infrastructures or leaving the infrastructure that is most comfortable behind is, quite simply, people. The emphasis on trust and respect demonstrated in Chapter 5 can be understood as a response to the uncertainty of the new trajectory a person is entering or aligning with. Identifying someone you can trust to help you orientate is as important for adult research participants as it is for a child at a new school. Without trust, disorientation increases, as individuals concerned can never be sure that the information they are receiving is a genuine representation of the new infrastructure or a strategic ploy to meet other ends. Essentially, even if you do not know the infrastructure of the partners you are planning to work with, if you can trust the person who does know those structures, life is a lot easier.

Conversely, if the infrastructure itself is not trusted, it becomes all the more challenging for individuals within those infrastructures to gain trust among others. CSIRO, with its reputation for 'aggressive', 'predatory' behaviour noted in Chapter 5, clearly had an organisational infrastructure that was quite different from that of the universities or government agencies, and one that many non-CSIRO partners had encountered before. Consequently anyone who operated primarily within the infrastructure that generated the types of activity that had earned those labels was, by default, not fully trusted until proven otherwise. Inter-personal contact allowed individuals to be separated from their infrastructures—for their own trajectories to be distinguished from that of their parent organisations.

In the case of the CRCs, constructing a new infrastructure likewise relied on trust that the construction process was fair and equitable, not being skewed towards others' existing trajectories. The case of Program 1 in the first iteration of the Greenhouse Accounting CRC was an example of this. As Program 1 was a basic research program almost exclusively conducted by a single research school at the Australian National University, there was some suspicion by others outside this Program that the CRC designers were 'feathering their own nest' rather than building a CRC on the basis of the best scientific approach. This suspicion was less damaging than it could have been, as the Program Leader was already known and trusted by many of the CRC partners. Over time, the acceptance of this program increased (it was reinstated in slightly altered form in the second Strategic Plan), in part as people came to know and trust the researchers involved.

The importance of personal contact in the development of both CRCs was vital, and it was a common refrain to hear that the face-to-face group meetings were what moved the CRC forward. As one researcher from the Coastal CRC stated it, reflecting on the meetings held in the early stages of his project:

> *even though I hated them [the meetings] at the time, looking back I would make them even longer. I'd allow much more time for people, because it's only after the first four or five days that you start to get any sort of feel for each other, whether you could work with them, what their skills were, what they were interested in.*

This was a tacit acknowledgement of the importance of personal contact that allowed people to better understand each others' trajectories, and better exploit the flexibilities within those individual trajectories to create a common direction and sense of momentum.

Talk of infrastructure and trajectories and timetables tends to hide the personal dimensions of doing the research in a framework that can easily sound overly structural and deterministic. However, as the last section highlighted, these concepts are fundamentally rooted in the notion of individuals making decisions and acting within and across the infrastructures that support them. The emphasis here is on *support*, not containment. Infrastructures support individual trajectories, they help to give them structure and meaning. Thus the large-scale scientific infrastructure, through defining what is and is not science does not draw a boundary around certain activities that includes some and excludes others (although the Science Wars noted in Chapter 2 can be viewed in that way). Instead, there is a process of mutual support—the individual supports the infrastructure by conforming to its structures and the infrastructure supports the person by providing a relatively stable context of meaning for that activity.

This is an important shift away from more common concepts of barriers and boundaries, as it reinstates the individual decision-maker as an active participant whose activity takes place *in relation to* the infrastructures that give it meaning. It is significant in the concept of integrated research, as the alignment of trajectories can then be understood as the conjunction of different sources of meaning. These can be of various scales; a CRC itself can provide a way of understanding research that is quite different from conventional science. They can also shift as people seek meaning in more than one infrastructure; the research that is contracted to inform a particular management decision (should the Council build more channels to flush mangroves or is it a waste of money?) can also be the subject of an academic paper (do constructed canals to flush mangrove lead to improved ecosystem health?), as different action contexts can be aligned to achieve complementary outcomes.

Infrastructures that cling...

The adoption by the Coastal CRC of Decision Frameworks and Citizen Science as the flagship themes of the CRC indicated a strategy to provide a relatively clear alternative to the infrastructure of conventional science, as opposed to the messy, unstructured collision zone. These were interesting approaches as they hovered at the edges of the science infrastructure, and provided structured pathways into other infrastructures. Decision Frameworks and Citizen Science not only exploited the gaps and interstices between science and the action–decision worlds, but shaped them, expanded them, and attempted to build solid bridges between them. These were undoubtedly bridges *from* the science infrastructure: academic databases to help researchers understand and do participatory research, Multi-Objective Decision Support Systems to support formalised participatory decision-making, media and leadership training for researchers, and so on. Yet several researchers and non-research participants were dubious about the efficacy of these bridges.

In terms of trajectories, the processes that were being proposed by these two Themes had particular characteristics that may help explain this reticence. One was their formality. Participation in formal processes locked people into particular timetables—their trajectories became bound together for the duration of the process. Committing to these processes meant reducing the individuals' flexibility and ability to adapt to change. Even where flexibility was built in, such as in the proposed MODSS process, this was flexibility restricted to certain parameters.

In other words, there was a perceived conflict between off-the-shelf partici-patory processes that used a generalised approach, and the sensitivities of working across trajectories based within a specific action context. Those who were already managing a careful weaving across a range of trajectories felt this most keenly. Their awareness of the subtleties and nuances of their own trajectories, as well as their knowledge of the trajectories of their partners, highlighted the risks of implementing a well-meaning but ill-suited formalised process.

This in part explains the situation in the CRC where many more researchers expressed an intent or preference to involve stakeholders through participatory processes than those who incorporated methods to actually do so. It tended to be assumed by those committed to increasing participatory processes across the CRC that such 'in principle support' was a reflection of lack of skills or lip service to the concept, and in some cases this may be justified. However, in other situations it was perhaps more the exercise of due caution, lest the shared trajectories which had already been carefully established be severed in the rush towards a more recognisable form of 'participatory research'. In this sense, parochial reluctance to bring in 'outsiders' is less a form of irrational insularity than an acknowledgement of the sensitivity of conducting relationships based on a specific action context.

CREATING NEW INFRASTRUCTURES

The CRCs themselves represented an effort to create a new infrastructure, a new set of rules and conventions that shape relationships and interactions. The infra-structure of the CRC program, by having articulated rules and requirements, also acted as an anchor of certainty in its insistence on research management based firmly in time. The limited seven- or fourteen-year life of the CRC encouraged researchers to plan their research according to limited time horizons; reporting requirements and reviews based on the achievement of milestones likewise.

In other words, the rules of the CRC Program created a mini-infrastructure that spanned major trajectories and was only partially permeable. This was advantageous in terms of reducing uncertainty; in many respects, by defining many of the criteria for 'success' of a CRC and insisting on Centres' accountability to those criteria, as discussed in Chapter 3, the CRC Program offered a relatively clear trajectory for achieving integrated research. Some of the devices used to create this trajectory were organisational guidelines and milestones.

Organisational guidelines

The range of guidelines that were proffered by the CRC Program and used by the CRCs themselves to guide the ways the Centres were designed and operated, as documented in Chapter 3, set some very clear rules of operation. First the *Guidelines for applicants* (CRC Program, 1999) played a very strong role in shaping how the Centres were originally proposed. Then the contract with the Government and the model contract suggested for use between partners, as well as the Second and Fifth Year Review Guidelines (CRC Program, 2001b, 2001a, respectively) all reduced the uncertainty of how to construct an organisation that might achieve integrated research and thereby make a difference. The program effectively offered a formula for achieving the goal of research that has influence on the 'real-world' action context of environmental management. As such, the CRC Program's guidelines and contracts formed a highly visible part of the CRC infrastructural matrix.

However, the 'carrot and stick' of funding was not the only force behind the Program's influence in how researchers interpreted the constraints and opportunities of heading down the path of a CRC trajectory. Because of the extensive reach of the CRC program into the Australian scientific community, the problems of any CRC 'in trouble' were rapidly aired throughout the community, with potentially damaging effects on the more senior researchers. For example, several researchers asked whether I had or would investigate the CRC for the Sustainable Development of Tropical Savannas. A bit like the black sheep relative whose exploits were interesting but a little embarrassing, this particular CRC was well known throughout the research community as having had major problems early in its life that required drastic steps (including the replacement of the CEO) and intervention by the CRC Program to repair. In other words, the CRCs were a relatively public undertaking, and while any major breakthroughs could turn into spectacular successes, problems were likewise widely broadcast. The historical experiences of this CRC were therefore well known as a possible trajectory if due care was not taken. These were more subtle, tacit parts of the CRC infrastructure that encouraged some conservatism in the organisational aspects of the CRCs.

Milestones

Another source of certainty at the project level was the use of milestones, commitments at the research project team level to achieve particular outcomes by specified times. As mentioned in Chapter 3, both CRCs used milestones extensively; every project that was funded through the CRC Program needed to have a full and agreed set of milestones, with accompanying deadlines.

These milestones reduced uncertainty into the future: whatever else is done, something must be done that looks enough like these milestones to be able to say they were fulfilled. In other words, the milestones themselves are not determinate, but rather yet another set of guidelines. As one researcher noted, there is something of an art to writing good milestones that sound firm enough to be recognisable as 'real achievement' yet vague enough to allow for a variety of trajectories that may unfold, allowing for calamities or unanticipated opportunities along the way.

The milestones themselves are also a source of self-induced pressure, which may or may not be productive. In the case of the SEQ Study projects that were contracted to the CRC, for example, the milestones and their deadlines were productive in the sense that they encouraged multi-disciplinary, multi-organi- sational teams to work together without allowing them the time to dwell on their differences. Simultaneously, similar pressure contributed to the rift between some of the researchers and the CRC, as the CRC was seen as hampering their efforts to meet their formal milestone commitments.

Milestones, then, to the extent that they shape the activity of integrated research, are powerful 'mini-infrastructures' in their own right, not laying out a specific path, but a set of boundaries within which the research must be carried out.

IS YOUR RESEARCH INTEGRATED? ☐ YES ☐ NO (PLEASE TICK ONE)

The theoretical framing used in this chapter has described integrated research as an ongoing process of ad hoc alignments, adaptive survival strategies and grappling for provisional new certainties through developing strong inter-personal relation- ships, locating oneself in specific action contexts, and adopting non-science timetables. Any, all, or none of these could serve as indicators of integration in any given instance. Yet any acknowledgement of the dynamic and processual nature of integration must also be an acknowledgement of its limitations as a 'tick-box' concept. This study has shown that, as a process, integration is always relational with respect to time, space and activity: integrated in comparison to when? Where? Which activities? Some activities will (it is hoped) remain much the same in integrated research—good laboratory practice, for example, or rigorous application of democratic processes of governance. However, for the purposes of assessment, some claims for integration often need to be made.

Representing integration

The question of how best to *represent* integrated research is central to this task. To continue to gain support, integrated research must be 'doing the job' and be seen to be 'doing the job well'. While some indicators such as the structure of an integrated research organisation, numbers of co-authored papers, public communications activities and extant formal linkages are currently used by the CRC Program as reported in the CRCs' annual reports and performance indicators, these are inevitably retrospective, and somewhat dubious as unambiguous representations of successful integration.

In this section I am not going to discard or even fully critique these attempts to pinpoint the moving target of integration. Instead, I will seek to identify those features that capture the spirit of integrated research by focusing on action into the future.

Organisational structure and guidelines The organisational structures and guidelines offered by the CRC Program have already been discussed in this Chapter as sources of certainty, but they were also important signposts to the future—and not only a future of continuing funding. The significance of these infrastructures was their guidance in structuring an organisation that could achieve integrated research. As both CRC cases illustrated, structure alone was no guarantee that integrated research would occur—indeed, questions of how to identify and measure integration were bound up in these guidelines, but were not answered within them. Instead the guides served as a proxy, as a structure that insisted on cooperation at some levels (the need for partners from different institutional sectors, for example) and created opportunities for it at others. As one researcher described the process of locating oneself within a CRC:

> *I don't think there's any way that any process of negotiation between all those different interests [among the partners] can be ideal for anybody. I mean, these things are always a shambles, finding a compromise that meets different people's aspirations and needs, including getting something that will attract the potential financial backers.*

In other words, while formulaic to an extent, the CRC formula was only indicative, not deterministic.

The CRC Program itself had to tread a careful line between being too prescriptive and thereby stifling the very creativity and flexibility they wished to encourage, and being too accommodating and not providing the structure that was necessary to direct prospective or current CRC partners towards productive cooperation. The variety of research and action contexts demanded flexibility from the Program as well as some certainty. One way they managed this process was through encouraging interaction among CRCs and the extensive use of exemplars, to highlight how others had achieved different goals through the CRC Program. Their primary vehicle for doing this was the annual CRC Association conference. The conference spent considerable time showcasing different aspects of CRC work across different sectors. This had the dual advantage of presenting good publicity in an open forum, as well as offering further guidance to other CRCs on ways to achieve the outcomes the Program considered appropriate, without being prescriptive. As such, both explicit and more tacit signposts to the future were employed at the organisational level.

Strategic plans Strategic (or business, forward, development) plans are now generally part and parcel of any organisation, whether in business, government, research or community. They too are clear signposts to the future, generated from within, rather than without. These plans give the organisations a framework within which to direct their own particular trajectories within (or across) the matrices within which they operate. Strategic plans, including mission statements and 'visions' are highly visible, public statements of an organisation's anticipated or desired trajectory. Each CRC had either a strategic or business plan, or both (business plans were required as part of the guidelines).

However, in integrated research the picture is complicated by the range of organisations involved—there are usually several strategic plans that need to be contended with in integrated research, and some more specific and strongly enforced than others. Some participants were quite explicit about this: their activities needed to be consistent with their employing organisation's strategic directions otherwise their participation in the CRC would be withdrawn by their management. CSIRO's external earning targets and the CRC's drain on their capacity to meet those targets was another example. The local catchment association in the Fitzroy area had their own strategic plan, which they hoped the Coastal CRC would consider in their next round of research planning.

Consequently, strategic plans and their many variations are important indicators of trajectories, but the plans themselves are not sufficient. Research managers planning integrated research also need to be aware of how rigorously those goals are to be applied, and whether there are any specific interpretations of the plans that may not be self-evident in the plans themselves. This task was usually the purview of the members of the Board for formal partners but, as in the case of the smaller non-core participants, such as the Fitzroy Basin Association, not all relevant parties were represented in this way.

CO-CONSTRUCTING INTEGRATING: DESIGN

While these signposts for the future are relatively well known and widely used, they do not necessarily offer much insight into how research can be designed to be integrative. They either rely on 'rules' handed down by a body with the power to ensure they are followed, or adaptations of standard business practice such as planning and project management. Even the CRC 'rules' do not dictate how the integration is actually achieved, apart from strongly encouraging a business-like structure with a Board and Managing Executive made up of representatives of different research partners. Within that fairly loose infrastructure, there are many more detailed decisions to be made if the research is going to generate an integrative process.

Using trajectories as a focal construct, some fairly generic, process-based observations can be made about how integrative research may happen.

Entry points

While it is common to think of integrated research as resulting from a slightly romantic, chance meeting over a beer in which two people serendipitously mention their areas of interest from which fascinating new research is generated, in practice this is a highly inefficient and unreliable basis for integrated research. Instead, integrated research design needs to create readily identifiable *entry points* that people outside the usual boundaries can both recognise and feel invited to take up.

Entry points are points at which it becomes easier for trajectories to entwine; they are avenues for participating. CRCs themselves can be described as base-line entry points, due to the partnering structure. For an entry point to be effective it needs to be widely recognised *as* an entry point, a manufactured gap in the infrastructure through which trajectories are invited to come into contact. Advisory councils and committees are exemplary entry points—an invitation to sit on such a committee is an invitation to participate in the large-scale research trajectory.

At the smaller scale, however, entry points *and their consequences* become far less clear. Can non-scientists legitimately participate in the 'scientific' research, or do they have to become part of the study in the sense of *being studied* (the 'let's get a social researcher in' scenario). Is data collected by a community-based group of volunteers of sufficient reliability to use as a basis for a scientific paper? Should contract research over which the research organisation has little control be allowed to consume significant proportions of available resources? These entry points in the cases of this study were in hot dispute, and there were no clear answers. Yet this poses a problem in terms of integration. If research bodies want to 'get people involved', what can they do with them once they are in? In other words, consideration does not stop with entry points. Any decision to create or participate in an entry point needs some structure to support those who enter. What happens to the trajectory next?

Momentum

The case of the Coastal CRC demonstrates the dangers of entry points that are not supported over time. Their efforts to generate enthusiasm and support for their original bid, and stakeholder workshops in the early days of the CRC were effectively efforts to highlight entry points, to invite people to join—possibly help shape—the CRC trajectory. There was reciprocal enthusiasm, and some momentum was built as stakeholders 'signed up'. The following pause in proceedings *relevant to the stakeholders* as the business processes were built allowed those trajectories to lose their momentum. The enthusiasm waned, and the CRC's reputation as being a research organisation that was genuinely interested in working with stakeholders (a reputation they had worked hard to create) was tarnished as people felt their time had been wasted and their paths misled.

As such, entry points alone will not generate a successful, integrative relationship over time. Some supporting infrastructure needs to be provided to ensure newcomers have ways of exerting their own influence on the trajectory of the CRC, and that the CRC can take an active role in participating in their futures.

Over-designing

If there are risks in under-designing integrative processes, there are also risks in over-designing. While the romantic view of integration was perhaps a little disparaged earlier, it nonetheless remains that much of the value of integration lies in allowing, encouraging, even forcing such interactions to occur. As mentioned earlier in this Chapter, a major potential benefit of the core sites concept for the Greenhouse Accounting CRC was that it provided a structure within which quasi-serendipitous insights were more likely to emerge. In other words, as noted with respect to the CRC Program infrastructure, at the organisational and project levels there is also a balance needed between creating an infrastructure that supports interaction and good relationships, and one that stifles it through lack of flexibility. One researcher suggested that the Greenhouse Accounting CRC had inadvertently done just this in committing to too many milestones in the bid process, leaving too little flexibility to follow emergent synergies or opportunities for change. This lack of flexibility was recognised by the Centre's management, and remedied by the new collaborative fund in its strategic plan. While some engineering is necessary to allow the creativity to happen, too much can stifle the space for innovation and newness, as trajectories become so tightly coupled that they cannot participate in any spaces that may be beneficial.

This raises further questions of how integrated research structures can be designed to cater for this variability in specific situations, rather than apply 'rules of thumb' that may or may not work in any given context. In other words, can integrated research itself be designed more *strategically* (in the sense of aiming for specific future scenarios) towards making a difference?

ORGANISING TO MAKE A DIFFERENCE: ZONES OF CHANGEABILITY

Taking the trajectories even further into the future introduces the 'acid test' question: how does all this happen in such a way that the research actually can make a difference? At this point, except under very prescribed circumstances (such as contract research to inform a specific policy decision) most analyses break down—indeed, there are few words or concepts to help here. Unless there are *actual* changes that can be identified, concepts of change are limited to such concepts as 'changing attitudes' or 'increasing awareness' or 'building relationships' and, of course, 'making a difference'—for what? While the other grey areas can be worked through more or less blindly in terms of getting integrated research processes underway, understanding where and how the partnerships can bring about change is crucial to the ideas of what integrated research is *for*.

Opportunities for change

As discussed extensively in Chapter 6, different research partners clearly have different fields of action, different capacities to bring about change; that is, after all, the point of attempting to bring them together. However, gaining and using an understanding of a partner's (or of several partners') action context can be a complex matter. In many instances throughout the analysis of these cases, sources of conflict and tension can be traced to inaccurate understandings of what the fields of action of different participants were—how their trajectories were restricted and where they were free to move; where the infrastructural matrix was tightly braided and not negotiable, and where it was openly woven and mutable. Looking forward over time, understanding how infrastructures and trajectories translate into opportunities for effecting change can enhance the research designer's or manager's ability to target some changes and avoid others.

Each individual trajectory operates within many such zones: my own research trajectory is shaped by my personal career aspirations; the organisational strictures and flexibilities of holding a research position; my historically-based domains of knowledge and expertise; my disciplinary affiliations; my networks of contacts; the communities I feel I am a member of; family obligations; political leanings; ethical beliefs, and many more. Some of these are more negotiable than others, some may be expanded or contracted more easily. As such my own personal trajectory and the constraints upon it can be described as a zone within which I can bring about change. I can revert to being an economist, I can seek out people who may support my research, I can forgo family life and spend more time in my office, I can join a political party and commit my work to supporting their aims. The further ahead I project from the present the larger the possible zone becomes, as more things become mutable over time. This is my own personal 'zone of changeability'.

Individuals who come together have zones that interact, and it is at this point that joint action and making a difference comes into play. I can offer to work with my cases to help them implement the ideas that emerged from this study, and thereby make a difference (hopefully) in their work practices, but it is beyond my own personal zone as to whether they would accept that offer. The potential for action that makes a difference, then, lies in the intersection of different zones of changeability. By bringing them together, each individual has the potential to co-opt the zones of others; others can co-opt that part of my zone of changeability that can adapt my existing knowledge to the scenario of management; I can co-opt their capacity to test the applicability of my ideas in a real-time situation.

Each potential integrative partner can be understood as operating within the context of an imagined zone of changeability, those degrees of flexibility they anticipate in their own trajectories as they extrapolate into the future. At the point of contact, these zones intersect, as illustrated in Figure 27. Integration allows people to *act into* zones that are not their own—research gains influence through adopting the influence of others.

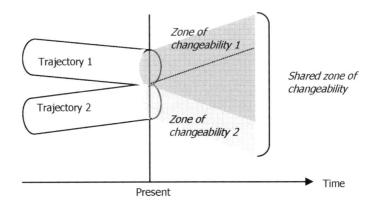

Figure 27. Zones of changeability.

While few would deny that maintaining good relationships is hard work, the concept of zones of changeability offers some sense of why this hard work may be worthwhile—and when it may be less than worthwhile. Work to understand how different partners can exert influence in the world may illuminate a wide range of strategic trade-offs useful for integrated research participants to consider. For example, the potentially most productive relationships (and it should be empha-sised that zones of changeability are all potential, as are future extrapolations) will change depending on the time horizon being analysed. For example, in the short term coastal research may be most influential by creating entry points and trajectory support for the local lord mayor; over the long term it may be more effective to entwine the trajectories of the national association of school teachers to incorporate coastal issues into the national high school curriculum. A balanced approach may do both. In the short, medium and long terms the greatest zone of changeability may be for senior greenhouse accounting scientists to participate directly through the entry points in the international climate change negotiations that will flow on to shaping national policy, rather than trying to create new entry points domestically.

In some respects, the concept of a zone of changeability is doing little more than giving an explicit label to judgements that are already made tacitly. However, a zone of changeability is, as I have tried to illustrate, an incredibly personal thing—it is a key defining feature of being an individual able to act within a particular context. Tacit judgements about other people's zones of changeability are therefore being made continually, sometimes quite erroneously, sometimes too conservatively, sometimes accurately but with additional guesses regarding their 'underlying motives'. Above-board discussions of how far people can comfortably move within their zones, where they would like to be able to go, and how they can enlist partners with the appropriate zone of changeability to help them get there could change the face of how integrated research is done, and to what extent it can make a difference.

Reverse planning

In this sense, a process that starts from the end, rather than from the beginning, asks very different questions. Rather than starting with questions of 'how can we get stakeholders on board?', planning starts with questions of 'what can I change?', 'what can we change together?', 'what do we want to be able to change?' and 'whose trajectory can we tap into to achieve that?' Moving backwards from this point, research designers can then strategically pinpoint who they need to approach. These may be people who are working strongly in the same location, such as the Lord Mayor of Brisbane and his public mandate to clean up the coastal waterways. They may be people who perhaps have not considered how coastal research can move into their own zone of changeability, such as the national association of school teachers. The risks of one trajectory as opposed to another can be weighed up—while more work may be needed in the latter case to generate interest and enthusiasm, the researchers may have greater influence here than over someone already moving in a trajectory with its own momentum.

Preliminary ideas of the ways the partners can work together can then be assessed, including how flexible the trajectory support will need to be. For those trajectories of less momentum, a more structured approach may be appropriate, whereas a trajectory with considerable momentum may require greater flexibility in some directions. Then, finally, consideration closer to the present may lead to the fashioning of appropriate and attractive entry points that will allow the imagined future to emerge in reality.

Planning in space and time

As such, to achieve integrated research requires self-conscious planning across space and time. Of course, all organisations do this: strategic plans with their nested program and project milestones are manifestations of planning for the future. However, incorporating consideration of zones of changeability renders strategic planning for integrated research quite different from its business or research counterparts. Most research managers, following the traditional model of science and its application after the research is completed and published, need only consider the effects of their research in terms of sources of funding available to them. While in an era of decreasing public funding and increasing private research funding this is arguably becoming a stronger driver of research programs than in former decades, it is nonetheless a broad consideration of context, not detailed planning to bring about change. At the other end of the spectrum, contract-based, purchaser–provider research is often very clearly placed within a detailed framework for change. However, the researchers themselves often have little say in what that framework actually is, unless they are high-level scientific advisers. Researchers operating in this environment need to consider their fellow researchers as competitors, rather than co-operators, unless temporary, strategic alliances need to be built to get a particular contract.

Cooperative, integrative research has a capacity to carve out its own space in future change that is as broad and as limited as the ability of its partners to bring about jointly desired differences. Yet freedom and influence are not without cost. Common ground in shared action contexts and the effects of research within those new boundaries can replace common institutionally-bound criteria for success. Becoming active contributors to localised political arenas gives researchers a political voice in contrast to their traditional role of impartiality and neutrality. Research programs that have been carefully crafted to fit with community and industry agendas likewise remove the veil of disinterestedness that is commonly viewed as a criterion of good science. In short, the markers of good science have changed. But changed to what? This question will form the basis of the concluding chapter.

CHAPTER 8

'HERE BE DRAGONS'

Negotiating the Future

We have no illusions about 'quick-fix' solutions. We have tried to point out some pathways to the future. But there is no substitute for the journey itself, and there is no alternative to the process by which we retain a capacity to respond to the experience it provides. (WCED, 1987/1990, p. 353)

As the participants in the study have demonstrated, there is currently no shortage of hidden dragons in the realm of integrated research. The shadows they hide behind include traditional understandings of science and poor conceptual links between those understandings and their participation in the dynamic, active world in which research brings about change. Increasingly, researchers and non-researchers alike are finding themselves confronting the shadows at the cusp of these two domains, feeling their way in the dark, stumbling at the mercy of half-seen forces. They are finding ways to articulate and negotiate their activity that are local, and grounded in their own histories and immediate experiences. Yet without broader conceptual frameworks to link their experiences to, the learning derived from those experiences about the cross-roads landscape tend to remain local, even individual. As this study has shown, through systematically exploring how people make sense of this landscape, it is possible to formulate empirically-based concepts for articulating some aspects of these experiences. These can serve as landmarks that may be identified in the shadows and can be negotiated.

From the outset this study has been concerned with understanding and articulating *processes* of integrated research. The ability to negotiate with both awareness of the nuances and subtleties of the situations that confront us and the vocabularies and conceptual structures that allow us to articulate that awareness, are vital to navigating the challenges of crossing entrenched boundaries. To re-state the main concern of this study, those aspects of our experience that we cannot communicate remain hidden, shadowy forces that are poorly comprehended, limited sources of learning, and inadequately planned for.

This study set out with four aims. One of these was to develop and apply a methodology that combined research as practice and social communication approaches as a basis for theory development that is relevant to both science studies and science. The concept of integrated research was built from conversations with people who were immersed in its practice, and by engaging those people in its construction through ongoing research conversations. These conversations formed a basis for constructing new ideas of what integrated research may be, how it is done, and how theoretical development in this area may contribute to its improvement.

In Chapter 4 I showed that there is a strong tendency to view integrated research as a rational process of manipulating flows of information, most commonly to achieve a more 'complete' picture of the world. In Chapter 5 we saw that despite this view, the actual work required to construct an integrated research program, to get it functioning and to participate in it, was not only scientific but also historical, political and social. These social factors were often spoken of as being more challenging than the technical ones, despite the science often being highly complex. The separation of the rational, scientific processes from the human, socio-political ones can be understood as a reflection of the traditional view that science should be abstract and decontextualised, independent of its social influences. These two chapters, and the tension between them, met the aim to develop a 'current' account of integrated environmental research 'from the trenches', based on the participants' own articulations and negotiations.

A third aim of the study was to contribute to the development of conceptual tools that can be used to better articulate the activities of doing integrated research, and thereby contribute to how others can learn from them. I began this development in Chapter 6 by shifting away from the experiences and stories of the participants as they had been observed and told, towards a more theoretical reconstruction of those experiences. By focusing on events where people bridged the self-imposed categorical divide between social and technical aspects of integrated research, alternative categories were highlighted. A theme that was common to the stories reconstructed in Chapter 6 was the relationship of the research to the context within which it would be turned to action, and how that action context could change over time. Chapter 7 developed these concepts further, focusing particularly on the temporal, forward-looking aspect of integrated research. Using the further theoretical layers of information infrastructures and trajectories as tools for temporal analysis, I re-told the stories of Chapter 6 from a temporal perspective. Current systems for designing, planning and assessing integrated research were reviewed from the perspective of integrated research being located in a dynamic action context. At a general level, I argued that the tools for planning and assessment that currently prevail cater only superficially for these dimensions of research.

This leaves the final aim: to contribute to the philosophical understanding of integrated environmental research, and perhaps, integrated research more generally, and how it differs from conventional research. Accordingly, this concluding chapter steps away from the minutiae of research practice and turns the analytical tools to science writ large. How can integrated research be understood in comparison to the predominant category (or categories) of traditional science? What might this mean for the trajectory of science as an institutional whole?

CATEGORISING SCIENCE: FOUR-DIMENSIONAL RESEARCH

Having journeyed through the complex terrain that both created and was created by the layers of intertwined practice and theory of the previous chapters, integrated research can now be simplified again into a more abstract schema. This schema

I shall call 'four-dimensional research'. In explaining these dimensions I will offer one way of understanding integrated research as specifically different from conventional science at the theoretical level.

So far I have argued that the traditional conceptual categories by which science and research are most commonly known (while sufficing for many research activities) are inadequate to understanding and articulating integrated research contexts. The dimensions presented here are a contribution to this articulation. However, these dimensions are not mutually exclusive categories, but are rather increasingly expanded views of research activity. By describing them as dimensions I intend to emphasise that they are not separate activities, but are (to return to an ongoing theme) different ways of looking at research practice, and what different categorical choices include and exclude.

Dimension one: science as pure thought

Although only fleetingly introduced here, one-dimensional science forms the first basic unit of this construct. One-dimensional science is the idea of science as 'pure thought', a disengaged, logical, rational process that underlies what makes science different from other modes of belief or inquiry.

While this understanding of science has come under increasing attack in recent decades, as outlined in Chapter 2, it still lies at the heart of a popular conception (popular both within science and in the general public) of what science is and 'how it ought to be'. It is inherent in a range of archetypal scientists: from the ancient 'Eureka!' example, in which Archimedes progressed ideas of specific gravity in his bath, to the more modern exemplar of Einstein solving the puzzles of the physical universe in the Patent Office and, more recently, the "Beautiful Mind" example, in which a socially-reclusive, clinically schizophrenic mathematician can win a Nobel Prize and have a successful Hollywood film based on his remarkable life. Each of these poses the essence of science and research as an isolated, individual process of pure thought, where context is remote and alienable.

While critics have tended to react to the romanticism of such portrayals of science, there can be little doubt that the work of highly technically skilled, creative individuals lies at the heart of research activity. However, while necessary it is, perhaps, too well celebrated, as this individual work is not sufficient for a person to be engaged in 'science'.

Dimension two: science as a social institution

Two-dimensional science builds on one-dimensional science: drawing on Newton's famous phrase "If I have seen further than most it is because I was standing on the shoulders of giants", it shifts attention towards science as a group activity. While isolation and individual pure thought may be the essence of science, in practice science is a collegiate process of building on the work of others. Thus science is viewed as a very large-scale group effort, in which processes such as publication, peer review and collaboration come to the fore. In two-dimensional science all players are, of course, scientists. This encompasses both notions of building upon the research that has already been done, by drawing on work published in academic journals, for example, as well as research performed as teams.

This view of science is also variously celebrated in popular scientific culture. Paradigmatic models here include the Manhattan Project to build the first atomic weapons, and even well known scientific fakes, such as cold fusion. Important in each is the characterisation of science as a *system* to generate reliable knowledge, where the knowledge is created through shared action. Sociological and historical studies that focus solely on the construction of knowledge within science explore this perspective of two-dimensional science. Interdisciplinarity and its variations live here. It is also where most of the models of Chapter 4 are located. This book, as a tool primarily designed for communication among researchers, exists in two-dimensional science.

Visually this can be represented as the flat plane of two-dimensional space, illustrated in Figure 28.

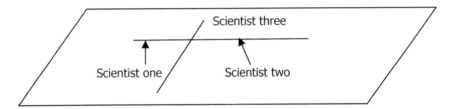

Figure 28. Two dimensional science

The plane metaphorically represents an ideal space within which science can operate without reference to the broader world. It is internally consistent and self-referential, essentially an independent scientific world.

Dimension three: enter the non-science players

In the third dimension, while pure thought and scientific cooperation remain, the social world enters consideration, as other planes intersect the two-dimensional world of science. This intersection does not negate the first two dimensions, but strengthens it and provides opportunities for research to move beyond the confines of science. This is the world of extension and the transfer of technology, as well as the world of the sociology of scientific knowledge, and many participatory models. It is the landscape of environmental law, and science communication, for example. Paradigmatic examples of three-dimensional science include the Montreal Protocol to protect the ozone layer and international bans on whaling. Local examples include Australia's National Carbon Accounting System, as well as feral pest control and measures to protect endangered species. Three-dimensional scientific space goes beyond where science happens to encompass the decision-making arenas where it is decided what science gets done, and what gets done with the science.

Theoretically, three-dimensional science is the territory of boundaries and categories, where different planes of action meet the scientific one. This is illustrated in Figure 29.

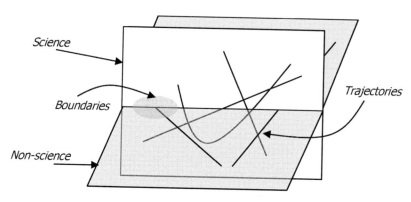

Figure 29. Three-dimensional science.

Boundaries and structures are surely present in three-dimensional research, but individuals can span them—perhaps not painlessly or without consequence, but the mobility of the individuals, and in particular their capacity to negotiate these points of intersection, enables research to be integrated across those planes. As the two previous chapters highlighted, one of the key points at which two planes of action intersect is place, the physical location of the research or the site of action.

It is a rich and exciting world, but it is not the end of the story.

Four-dimensional research

The fourth dimension is, unsurprisingly perhaps, time. Four-dimensional research is carried out in connection across researchers and with non-research communities over time. It is in four-dimensional space that research can influence action, that it can make a difference. Boundaries and categories exist, but they are not static—they exert pressure and torque against each other, influencing the directions and speeds of the joint action.

Where are the theories and exemplars of four-dimensional research? As yet they are few and far between, not because the practice of four-dimensional research is rare—any research *can* be seen as four-dimensional. However, it is not often recognised as such. The models of Chapter 4 indicate that the work that goes into maintaining good relationships over time is not commonly seen to be work; understanding where your research partners are heading and the constraints they are operating under is usually regarded as common sense or necessity, not a complex interweaving of action contexts and how they can be extrapolated into the future. The sense of how far a project leader can push a team researcher to spend more time on the joint project and less time on their own pet project is seen as (usually tiresome) management, not a well-developed sense of who the team member is, how she relates to the joint project and to her own home organisation, where she sees her career headed, and the strength of the leader's relationship with her. Judgements regarding how closely research should adhere to current political agendas and how much they should try to lead them in new directions are not only judgements of the here-and-now, but assessments of the past, present and future of both the politics and the science.

In terms of integrated research, then, it can be suggested that to be self-conscious about how research might bring about change means to be incorporating an awareness of the third and fourth dimensions of research. This awareness need not be explicit (the cases have illustrated that it rarely is), but it needs to be present in the design, planning and conduct of research. This brings us to the question of what the explicit consideration of four-dimensional research might entail.

What might four-dimensional research look like? The first dimension requires that, like any research, it displays some degree of originality and creativity, as well as the application of sound research principles and methodology. The second dimension requires communication and rigour that will pass the intellectual standards of the research community. These criteria define an activity as 'research' rather than as casual investigation. The third dimension requires that the research actively incorporate an understanding of the action contexts within which the research is situated, and places the research in contact with at least some of those contexts—that is, it would need to be able to demonstrate that it was operating at the intersection of the planes of science and another institution. Finally, to incorporate the fourth dimension, it would need to allow for, or anticipate, changes in the relationships between research and the action context over time.

Some existing research models meet these criteria. Some of the more recent adaptive management approaches, mentioned in Chapter 2, do attempt to incorporate social contexts of research, and their emphasis on experimental, iterative management brings some aspects of change into the research process. Some systems approaches, particularly critical systems thinking (which includes ongoing reflexivity with respect to the role of researchers in the systems they are researching, concerns with emancipation, and research across disciplines), can also be classified as four-dimensional (for an introduction see Midgley, 1996). In research with a more social orientation, participatory action research, also mentioned briefly in Chapter 2, and some of its variants in community development research also emphasise social context and processes of change, may also be examples of four-dimensional research. This study, and the methodology of social communicative practice, by marrying the technical dimensions of rigorous design, sampling and interview procedures with a concern for building good social relationships and engaging in an ongoing conversational process, was also (in retrospect!) four-dimensional.

While these methods and methodologies each incorporate aspects of four-dimensional research, they have not been theorised in this way. What might an overarching structure such as four-dimensional research add to this already complex research environment?

WHY THEORISE SCIENCE IN FOUR DIMENSIONS?

This question returns us to the beginning of this story, and the issue of choice of integrated research methods. If researchers intend to have their research make a difference, how does one choose between an adaptive management approach and a participatory action research methodology, for example? Is it commonly recognised that these two very different research methodologies (and others) can serve similar purposes, in terms of linking research and action? Currently, as this study has shown, these questions are rarely addressed in a systematic way, if they are recognised explicitly at all. The concepts that were presented in previous chapters offer one way of bringing the options into the conversations of integrated research, considering them openly and transparently, with some tools to begin to assess which may be appropriate to different circumstances. However, while concepts of action contexts, trajectories and infrastructures, torque and twisting may offer rich insights into temporal processes, they represent only a modest start for a concept so fundamental to understanding the processes of four-dimensional research. New concepts, such as zones of changeability, may increase our capacity to identify and analyse issues concerned with dynamic action contexts, but they too are raw and largely untested.

Of course science as an activity takes place in four dimensions—even hard-line subscribers to the one-dimensional model of science would be hard pressed to deny that science takes place embedded in a wide range of contexts over time. Yet it may be speculated that the dominant paradigm of science (the one- or two-dimensional models) still holds sway over perceptions of how science *should* be done. Even for those heavily involved in integrated research, the best science is still widely held to be that of the breakthrough, the next leap forward in knowledge and understanding, and that is reward in itself. Different approaches to research are most often regarded as deviations from this path, detrimental distractions from the single-minded pursuit of the next big thing.

Accordingly, even within the cases studied here, there was commonly a reluctance to bring the third and fourth dimensions of research *into* the realm of science—even while social and temporal factors were being actively incorporated into research planning, there was no conceptual alternative to the traditional division of science and not-science. Consequently, where people *did* want to understand the processes of integrated research better, they were confronted with mirages and veils rather than clarity and openness. It should be emphasised here that this was not the product of deliberate deceit or obstinate denial. It was the result of far more subtle processes, of not having ways to speak about the third and fourth dimensions of research without denigrating the science, of operating in two different research systems at once that rewarded different things, of unwillingness to replace the certainty of conventional research with the uncertainty of whatever good integrated research might be.

The construct of four-dimensional research offers a different way of thinking about integrated research that is not in opposition to traditional science. Rather, it represents the more formal incorporation of aspects of science *that are there anyway*, but are most often hidden. This is an important step towards understanding and articulating how integrated environmental research relates to conventional science that may open up new opportunities in planning and managing integrated research processes.

Traditional science infrastructure impinges

It would be simplistic to suggest that articulation is a panacea for integrated research. One of the most pervasive challenges for integrated research lies in attempting to do four-dimensional research in a research system or infrastructure (used in the common sense here) that is designed predominantly for justification and reward in one or two dimensions. In other words, the need to design for research in four dimensions does not stop at the organisational scale—organisations such as CRCs are constrained by the greater systems within which they operate. Important here, though, is the plural. Integrated research organisations need to fit within several infrastructures, not just a scientific one.

As noted earlier, while the CRC structure was occasionally constraining, it did provide a formula that people could work to, to build an integrated research organisation. While several researchers and research managers complained of the bureaucratic nature of CRC work, others who had experience of trying to generate research that crossed boundaries *without* having such structures to support them, were very glad of their presence and function. The CRC Program walked the line between guidance and freedom, which was adaptable to many situations, albeit perhaps never perfect for any particular given research context. The environment sector CRCs who participated in this study each, in vastly different ways, pushed the limits of integration in their design, planning and implementation. From creating an entire Centre tailored to meet the needs of specific Articles of the Kyoto Protocol to involving researchers in a wide range of community, government and industry contexts, both CRCs were engaged in a radical reconception of what it means to do environmental research in the 21st century.

However, one of the reasons such a traditional infrastructure can remain in force is because there are few conceptual tools with which to expand it. The strategies adopted by the CRC were steeped in risk and challenge, not least because of the vagueness with which they could justify their risk-taking. There was a sense of pressure to engage in four-dimensional research, while still being assessed—both internally and externally—predominantly under the criteria of one- or two-dimensional research. While the CRC Program, through its insistence on the involvement of research users, opened some doors to bringing the third dimension into consideration, the fourth dimension was absent. The concepts presented here will hopefully offer some ways of contextualising the risks of integrated research according to the ways in which researchers seek to make a difference, and thereby offer a basis from which their relationship to conventional science can be negotiated.

THE TRAJECTORIES OF SCIENCE: THE FATE OF THE CRCS AND SPECULATIONS ON THE FUTURE

As I discussed in Chapter 7, science as a whole can be understood as a massive trajectory, and it is worth considering at the close of this book how integrated research may be affecting the trajectory of science as a whole. Earlier, in Chapter 2 I also noted that the shifts towards integrated environmental research are driven by two major forces: the knowledge economy, and the democratisation of knowledge. While these two forces are currently pushing science generally towards greater integration, there is clearly a great conceptual, practical and moral schism between them. The fate of these two case studies (beyond the life of this study) has been caught up in the struggle between them, and the implications of this for integrated research and science as a whole are significant.

The knowledge economy trajectory is firmly rooted in the values of the neo-classical scientific paradigm: it aims to bring researchers and others together for the purposes of greater efficiency, greater financial returns, and ultimately greater economic growth. Politically it enhances the positions of the entrenched power elites, large corporations and technically-oriented companies, as well as global technological powers.

The knowledge democracy, in contrast, seeks to bring the knowledge of the disempowered into decision-making systems. By demolishing the arguments that support the privileging of science, and creating new philosophies in which the local knowledge of the disempowered becomes heard and counted, the knowledge democracy also seeks to demolish the very power elites that the knowledge economy is strengthening. Simultaneously, the independence of research and its ability to exert influence over already powerful institutions is also perceived to be threatened, and the knowledge autocracy is defending its claim to disinterested, expert status.

Integrated research is one place where these three trajectories come together, in the involvement of powerful corporations alongside community groups, the inclusion of conservation organisation representatives alongside land developers and oil company executives. The role of the researcher is increasingly one of mediation between these competing political and ethical positions, where the very mediation itself splits the research community into democratic and autocratic ideologies.

Can the science trajectory sustain such schism?

Science Wars Mark II?

The CRCs that featured in this study, and offered such rich insights into the processes of integrated research, approached the end of their first 7 year funding cycle in 2006. To gain a second 7 year funding agreement, they would have needed to reapply in 2004. Prior to the 2004 funding round, there were rumours that the Australian Federal Government was going to change the rules. These rule changes were unveiled in the Call for Applications—the CRC Program would no longer support Centres that could not demonstrate significant commercial or industrial benefit. From 2004 all Centres had to demonstrate a commitment to comercialisation as their core activity. These changes applied both to new Centres, and to existing ones applying for refunding. With primarily public good mandates, the Greenhouse Accounting CRC and the Coastal CRC were not eligible for further funding. They ceased operations in July and June 2006 respectively.

The knowledge economy staged a bloodless coup over the knowledge democracy—and there were very few shots fired in protest.

While all future trajectory extrapolations are speculative, it can be argued that the tension between the knowledge democracy and the knowledge economy is currently simmering below the surface of the scientific institution. This situation is reminiscent of the conflict between economy and democracy that was simmering below the surface of the political institution prior to the anti-globalization protests at the World Trade Organization meeting in Seattle in 2000. Is this tension likely to lead to the 'Integrated Science Wars'? After all, the current version of the Science Wars revolves around a conflict of values and beliefs that are perhaps no more dramatic or intense than this. Will the science trajectory be able to maintain a single momentum despite these two opposing moral forces, or will it split into two (or more) separate trajectories that work in different integrative contexts? Or will the two trajectories entwine in acrimony?

While the fate of the CRCs embodies the former, I believe there are indications that the latter can be regarded as a real possibility. As more research becomes 'integrated' with conflicting interests and conflicting moral values, so too is it potentially an arena of battle. Further, it is in the practice of integrated research in public good areas, such as environment, some aspects of agriculture and health, which will show—is already showing—the first signs of this pressure. Bringing non-research partners, their knowledge and their needs and values to an integrated research situation is, at some levels, a key advantage of integrated research—some might say it is the whole point. Yet at the individual and inter-personal levels, this can lead to turmoil. Some participants in this study showed resentment towards either or both forces: resentment at being 'told what to do' by non-scientists, and resentment at being at the 'beck and call' of industry or government to serve ends they may not aspire to. In other words, perceptions that the political values come to dominate the scientific ones lead to unrest and tension.

Regardless of whether researchers see themselves as value-neutral, impartial scientists or as active moral agents, as integrated research brings in partners with different values systems, eventually there will be conflict, either as these systems impinge on the researcher's self image of objectivity, or as moral values collide.

There can be little doubt, given the history of Western science over the last three or four centuries, that research will continue to play a highly influential role in the future of society. It is unlikely that demand from democratic and economic forces for increasing participation in science will diminish in the foreseeable future, as science and technology become increasingly embedded in all aspects of human understanding and activity. We do indeed face a socio-scientific future, and integrated research is one response to that realisation.

As always, the greatest threats that confront the existing institutions of society are also its greatest opportunities. The changing relationships between science and society that are embodied in integrated environmental research can be viewed in both ways. As a threat, it can undermine the independence and objectivity scientific research is founded upon. As an opportunity, it allows scientists and researchers to have direct influence in decisions that affect the biosphere and our ability to live within it. The better that society as a whole—including, of course, scientists—can articulate, understand and negotiate the dynamics of the relationships between science and society, the better equipped we will be to work out how we might achieve the futures we desire.

LIST OF REFERENCES

Ampt, P. R., & Ison, R. (1989). Rapid rural appraisal for the identification of grassland research problems. *XIV International Grassland Congress Association Proceedings*. Versailles, Nice: Française pour la Production Fourragere.

Anon. (1998). *Coastal zone, estuary and waterway management: A proposal.* Brisbane: CRC Australia.

Argyris, C., & Schön, D. A. (1996). Organizational learning II: Theory, method and practice. In E. H. Schein & R. Beckhard (Eds.), *Addison-Wesley series on organizational development.* Reading, MA: Addison-Wesley Publishing Company.

Ashman, K. M., & Baringer, P. S. (Eds.). (2001). *After the science wars.* London: Routledge.

Australian Government Solicitor. (1998). *CRC commonwealth agreement.* Barton: ACT.

Ausubel, J. H., & Langford, H. D. (Eds.). (1997). *Technological trajectories and the human environment.* Washington DC: National Academy Press.

Bammer, G. (2005). Integration and implementation sciences: Building a new specialization. *Ecology and Society, 10*(2), 6. Retrieved from http://www.ecologyandsociety.org/vol10/iss2/art6/

Barber, B., & Hirsch, W. (1962). *The sociology of science.* New York: Free Press.

Barnes, B. (1977). *Interests and the growth of knowledge.* London: Routledge & Kegan Paul.

Barnes, B., & Edge, D. (1982). General introduction. In B. Barnes & D. Edge, (Eds.), *Science in context: Readings in the sociology of science* (pp. 1–12). Milton Keynes: Open University Press.

Barnes, D., & Bloor, B. (1981). Relativism, rationalism and the sociology of knowledge. In M. Hollis & S. Lukes (Eds.), *Rationality and relativism.* Oxford: Blackwell Publishers.

Barr, N. F., & Cary, J. W. (1992). *Greening a brown land: The Australian search for sustainable land use.* South Melbourne: Macmillan.

Batterham, R. (2000). *The chance to change: Final report by the Chief Scientist.* Canberra: Commonwealth of Australia.

Bellamy, J., Ross, H., Ewing, S., & Meppem, T. (2002). *Integrated catchment management: Learning from the Australian experience for the Murray-Darling Basin. Final Report.* Canberra: Murray-Darling Basin Commission.

Bellamy, J. A., MacDonald, G. T., Syme, G. J., & Butterworth, J. E. (1999). Evaluating integrated resource management. *Society & Natural Resources, 12*(4), 337–353.

Board on Sustainable Development Policy Division National Research Council. (1999). *Our common journey: A transition toward sustainability.* Washington, D.C.: National Academy Press.

Bosch, O. J. H., Ross, A. H., et al. (2003). Integrating science and management through collaborative learning and better information management. *Systems Research and Behavioural Sciences, 20*(2), 107–118.

Bowker, G. C., & Star, S. L. (1999). Sorting things out: Classification and its consequences. In W. E. Bijker, W. B. Carlson & T. Pinch (Eds.), *Inside technology series*. Cambridge, MA: MIT Press.

Bradshaw, G. A., & Bekoff, M. (2001). Ecology and social responsibility: The re-embodiment of science. *TRENDS in Ecology & Evolution, 16*(8), 460–465.

Brannon, P. M. (2002). Our land grant mission in the twenty-first century. *Human Ecology, 30*(1), 1.

Brennan, M. (2001, May 15–17). *What constitutes a successful CRC application?* CRC Association conference: R&D return on investment, Perth, Western Australia. Unpublished proceedings.

Brown, J. (2001). *Who rules in science: An opinionated guide to the wars*. Boston: Harvard University Press.

Bryman, A. (1989) Research methods and organization studies. In M. Bulmer (Ed.), *Contemporary social research series*. London& New York: Routledge.

Burns, T., & Stalker, G. M. (1961). *The management of innovation*. London: Tavistock Publishers.

Buttel, F. H. (1992). Environmentalization: Origins, processes, and implications for rural social change. *Rural Sociology, 57*(1), 1–29.

Byatt, I. C. R., & Cohen, A. V. (1969). *An attempt to quantify the economic benefits of scientific research. Science policy studies series no. 4*. London: Hmso.

Cernea, M. M., & World Bank. (1991). Putting people first: Sociological variables in rural development. New York: Oxford University Press.

Chaiklin, S., & Lave, J. (Eds.). (1993). Understanding practice: Perspectives on activity and context. In R. Pea & J. Seely Brown (Eds.), *Learning in doing series*. Cambridge: Cambridge University Press.

Chalmers, A. F. (1982). *What is this thing called science? An assessment of the nature and status of science and its methods*. St Lucia, Queensland: University of Queensland Press.

Chambers, R. (1980). *Rapid rural appraisal: Rationale and repertoire*. Brighton UK: Institute of Development Studies at the University of Sussex.

Chambers, R. (1983). *Rural development: Putting the last first*. London: Longman.

Chambers, R. (1997). *Whose reality counts?: Putting the first last*. London: Intermediate technology Publications.

Chambers, R., & Jiggins, J. (1986). *Agricultural research for resource poor farmers: A parsimonious paradigm*. Brighton, UK: Institute of Development Studies, University of Sussex.

Chambers, R., Pacey, A., & Thrupp, L. A. (1989). *Farmer first: Farmer innovation and agricultural research*. London: Intermediate Technology Publications.

[Coastal CRC, see CRC for Coastal Zone, Estuary and Waterway Management]

Commonwealth of Australia. (1992). *National strategy for ecologically sustainable development*. Prepared by the Ecologically Sustainable Development Steering Committee, Canberra.

Conway, G. R. (1987). The properties of agroecosystems. *Agricultural Systems, 24,* 95–117.

Cooke, B., & Kothari, U. (Eds.). (2001). *Participation: The new tyranny?* London: Zed Books.

Cortner, H. J. (2000). Making science relevant to environmental policy. *Environmental Science and Policy, 3,* 21–30.

CRC Association. (2000). *About CRCs.* Retrieved May 29, 2000, from http://www.crca.asn.au/

CRC for Coastal Zone, Estuary and Waterway Management. (2000). *Annual report.* Indooroopilly, Queensland.

CRC for Coastal Zone, Estuary and Waterway Management. (2001). *Annual report.* Indooroopilly, Queensland.

CRC for Greenhouse Accounting. (2000). *Annual report.* Canberra.

CRC for Greenhouse Accounting. (2001a). *Annual report.* Canberra.

CRC for Greenhouse Accounting. (2001b). *Strategic research plan 2001–2004.* Canberra.

CRC Program. (1999). *Guidelines for applicants: 2000 selection round and general principles for centre operations.* Canberra: Department of Industry, Science and Tourism.

CRC Program. (2000). *CRC Compendium.* Canberra: AusIndustry/Department of Industry, Science and Resources.

CRC Program. (2001a). *Fifth year review guidelines.* Canberra: AusIndustry/ Department of Industry, Science and Resources.

CRC Program. (2001b). *Second year review guidelines.* Canberra: AusIndustry/ Department of Industry, Science and Resources.

CRC Program. (2001c). *Model agreement for the establishment and operation of a Cooperative Research Centre.* Canberra: AusIndustry/Department of Industry, Science and Resources.

CRC Program. (2002). *CRC Compendium.* Canberra: Department of Education, Science and Training.

CSIRO. (2000). *CSIRO strategic plan 2000–2003.*

Davenport, S., Leitch, S., & Rip, A. (2003). The 'user' in research funding negotiation processes. *Science and Public Policy, 30,* 239–250.

Davidoff, F., DeAngelis, C. D., Drazen, J. M., & Hoey, J. (2001). Sponsorship, authorship, and accountability. *The Lancet, 358*(9285), 854–856.

Dosi, G. (1988). Sources, procedures and microeconomic effects of innovation. *Journal of Economic Literature, 26,* 1120–1171.

Dovers, S. R., & Mobbs, C. D. (1997). An alluring prospect? Ecology, and the requirements of adaptive management. In N. Klomp & I. Lunt (Eds.), *Frontiers in ecology: Building the links* (pp. 39–52). Oxford: Elsevier Science.

Doz, Y., & Hamel, G. (1998). *The alliance advantage: The art of creating value through partnering*. Boston: Harvard Business School Press.

Ernø-Kjølhede, E., Husted, K., Mønsted, M., & Wenneberg, S. B. (2001). Managing university research in the triple helix. *Science and Public Policy, 28*(1), 49–55.

Etzkowitz, H., & Leydesdorff, L. (Eds.). (1997). *Universities and the global knowledge economy: A triple helix of university-industry-government relations.* London: Pinter Publishers.

Etzkowitz, H., & Leydesdorff, L. (2000). The dynamics of innovation: From National Systems and "Mode 2" to a Triple Helix of university–industry–government relations. *Research Policy, 29*, 109–123.

Ezzy, D. (2002). *Qualitative analysis: Practice and innovation*. Sydney: Allen & Unwin.

Fals-Borda, O., & Rahman, M. A. (1991). *Action and knowledge: Breaking the monopoly with participatory action research.* New York, London: Apex Press, Intermediate Technology Publications.

Feyerabend, P. K. (1962). Explanation, reduction and empiricism. In H. Feigel & G. Maxwell (Eds.), *Scientific explanation, space and time* (Vol. 3, pp. 28–97). Minneapolis: University of Minnesota Press.

Feyerabend, P. K. (1964). Realism and instrumentalism. In M. Bunge (Ed.), *The critical approach to science and philosophy* (pp. 280–308). New York: Free Press.

Ford, P. (2001, July 24–26). *Fitzroy river water quality*. Poster presented at the Coastal CRC Annual Workshop, Noosa.

Foucault, M. (1966). *Les Mots Et Les Choses: Une Archeologie Des Science Humaines*. Gallimard, Paris: Bibliotheque Des Science Humaines.

Foucault, M. (1973). *The order of things: An archaeology of the human sciences.* New York: Vintage Books.

Foucault, M. (1976). *The archaeology of knowledge*. New York: Harper & Row.

Foucault, M., & Rabinow, P. (1997). Ethics: Subjectivity and truth. In M. Foucault (Ed.), *Dits et écrits* (Vol. 1). New York: New Press.

Freeman, C. (1987). *Technology policy and economic performance: Lessons from Japan*. London: Frances Pinter Publishers.

Freire, P. (1996). *Pedagogy of the oppressed*. Penguin, London: Penguin Education.

Fuller, S. (2000). *Thomas Kuhn: A philosophical history for our times*. Chicago: University of Chicago Press.

Funtowicz, S. O., & Ravetz, J. R. (1993). Science for the post-normal age. *Futures, 25*, 739–755.

Gaylord, B. (2000). *American accents in Australian executive suites*. New York: New York Times.

Gibbons, M., Limoges, C., Nowotny, H., Schwartzman, S., Scott, P., & Trow, M. (1994). *The new production of knowledge: The dynamics of science and research in contemporary societies*. London: Sage.

Gieryn, T. F. (1999). *Cultural boundaries of science: Credibility on the line.* Chicago: University of Chicago Press.

Glaeser, B. (1987). *The Green Revolution revisited: Critique and alternatives.* London, Boston: Unwin Hyman.

[Greenhouse Accounting CRC, see CRC for Greenhouse Accounting]

Gross, P. R., Levitt, N., & Lewis, M. W. (1996). *The flight from science and reason. Annals of the New York Academy of Sciences.* New York: The New York Academy of Sciences.

Gunderson, L. H., Holling, C., & Light, S. S. (1995). *Barriers and bridges to the renewal of ecosystems and institutions.* New York: Columbia University Press.

Habermas, J. (1963). *Theorie und Praxis: Sozialphilosophische Studien. Politica: Abhandlungen und Texte zur politischer.* Wissenschaft, Luchterhand, Neuwied am Rhein.

Hall, P. (1994). *Innovation, economics and evolution: Theoretical perspectives on changing technology in economic systems.* Hemel Hempstead: Harvester Wheatsheaf.

Harbison, J., & Pikar, P. (1998). *Smart alliances: A practical guide to repeatable success.* San Francisco: Jossey Bass Publishers.

Hinchcliffe, F., Thompson, J., Pretty, J., Guijit, I., & Shah, P. (Eds.). (1999). *Fertile ground: the impacts of participatory watershed management.* London: Intermediate Technologies Publishers.

Holling, C. S. (1998). Two cultures of ecology. *Conservation Ecology, 2*(2), 4.

Holling, C. S. (1978). *Adaptive environmental assessment and management. United Nations Environmental Program.* Chichester: Wiley.

Houghton, J. T., Ding, Y., Griggs, D. J., Noguer, M., van der Linden, P. J., & Xiaosu, D. (2001). *Climate change 2001: The scientific basis.* Contribution of Working Group I to the Third Assessment Report of the Intergovernmental Panel on Climate Change (IPCC). Cambridge UK: Cambridge University Press.

Huber, P. W. (1991). *Galileo's revenge: Junk science in the courtroom.* New York: Basic Books.

Irwin, A. (1995). *Citizen science: A study of people, expertise and sustainable development.* London: Routledge.

Ison, R., & Russell, D. (2000). *Agricultural extension and rural development: Breaking out of traditions.* Cambridge, New York: Cambridge University Press.

[IUCN] World Conservation Union. (2002). *The IUCN red list of threatened species.* Retrieved June 12, 2002, from http://www.redlist.org/info/ introduction.html

Jasanoff, S. (1990). *The fifth branch: Science advisers as policymakers.* Cambridge, MA: Harvard University Press.

Jasanoff, S. (1998). Coming of age in science and technology studies. *Science Communication, 20*(1), 91–98.

Kates, R. W., Clark, W. C., Corell, R., Hall, M., Jaeger, C. C., Lowe, I., et al. (2001). Sustainability science. *Science, 292*(April 27), 641–642.

Keen, M. (1997). Catalysts for change: The emerging role of participatory research in land management. *The Environmentalist, 17*, 87–96.

Kim, K. M. (1994). Explaining scientific consensus: The case of Mendelian genetics. In S. Fuller (Ed.), *Conduct of science series*. New York & London: The Guildford Press.

Klein, J. T. (1990). *Interdisciplinarity: History, theory, and practice*. Detroit: Wayne State University Press.

Knorr-Cetina, K. (1999). *Epistemic cultures: How the sciences make knowledge*. Cambridge, MA: Harvard University Press.

Knorr-Cetina, K., & Mulkay, M. J. (1983). *Science observed: Perspectives on the social study of science*. London: Sage.

Kuhn, T. S. (1962/1970). The structure of scientific revolutions. In O. Neurath (Ed.), *Foundations of the Unity of science series* (Vol. 1(2), 2nd ed.) International Encyclopedia of Unified Science. Chicago: University of Chicago Press.

Kuhn, T. S. (1977). *The essential tension: Selected studies in scientific tradition and change*. Chicago: University of Chicago Press.

Land and Water Australia. (2001). *Strategic R&D plan 2001–2006*. Canberra: Land and Water Australia.

Latour, B. (1987). *Science in action: How to follow scientists and engineers through society*. Cambridge, MA: Harvard University Press.

Latour, B. (1999). *Pandora's hope: Essays on the reality of science studies*. Cambridge, MA: Harvard University Press.

Latour, B., & Woolgar, S. (1979). Laboratory life: The social construction of scientific facts. In *Sage library of social research* (Vol. 80). Beverly Hills, CA: Sage Publications.

Lave, J. (1993). Introduction. Understanding practice: Perspectives on activity and context. In S. Chaiklin, J. Lave, R. Pea, & J. Seely Brown (Eds.), *Learning in doing series* (pp. 3–32). Cambridge: Cambridge University Press.

Lave, J. (1996). The savagery of the domestic mind. In L. Nader (Ed.), *Naked science: Anthropological inquiry into boundaries, power and knowledge*. London: Routledge.

Lave, J., & Wenger, E. (1991). Situated learning: Legitimate peripheral participation. In R. Pea & J. S. Brown (Eds.), *Learning in doing series: Social, cognitive and computational perspectives*. Cambridge: Cambridge University Press.

Lee, K. (1993). *Compass and gyroscope: Integrating science and politics for the environment*. Washington DC: Island Press.

Leeds-Hurwitz, W. (Ed.). (1995). *Social approaches to communication*. New York: Guilford Press.

Liebowitz, J. (Ed.). (1999). *Knowledge management handbook*. CRC Press.

Lubchenco, J. (1998). Entering the century of the environment: A new social contract for science. *Science, 279*(23 January), 491–497.

Lundvall, B. Å. (1992). *National systems of innovation: Towards a theory of innovation and interactive learning*. London: Pinter Publishers.

Margerum, R. D., & Born, S. M. (1995). Integrated environmental management: Moving from theory to practice. *Journal of Environmental Planning and Management, 38*(3), 371–391.

Mercer, D., & Stocker, J. (1998). *Review of greater commercialisation and self funding in the Cooperative Research Centres Program.* Canberra: Commonwealth of Australia, Department of Industry, Science and Tourism.

Merton, R. K. (1938). Science and the social order. *Philosophy of Science, 5*(3), 321–337.

Midgley, G. (1996). What is this thing called CST? In R. Flood & N. Romm (Eds.), *Critical systems thinking: Current research and practice.* New York: Plenum Press.

Mitchell, B. (1987). *A comprehensive-integrated approach for water and land management.* Occasional Paper No. 1. Armidale: University of New England.

Moreton Bay Catchment Water Quality Management Strategy Team. (1998). *The crew member's guide to the health of our waterways: Ecological health and water quality management in the Moreton Bay catchment—Queensland, Australia.* Brisbane, Queensland: Moreton Bay Catchment Water Quality Management Strategy Team.

Nelkin, D., & Lindee, M. S. (1995). *The DNA mystique: The gene as a cultural icon.* New York: Freeman.

Nowotny, H., Gibbons, M., Scott, P., & Cambridge, E. P. (2001). *Re-thinking science: Knowledge and the public in an age of uncertainty.* Cambridge, England: Polity.

OECD. (1996). *The knowledge-based economy.* Paris: Organization for Economic Co-operation and Development.

[OECD] Organization for Economic Co-operation and Development. (1999). *The management of science systems.* Paris: OECD.

Okali, C., Sumberg, J., & Farrington, J. (1994). *Farmer participatory research: Rhetoric and reality.* London: Intermediate Technology Development Group Publishing.

Penman, R. (2000). *Reconstructing communicating: Looking to a future. Lea's Communication Series.* Mahwah NJ, London: Lawrence Erlbaum and Associates.

Pickering, A. (1992). *Science as practice and culture.* Chicago: University of Chicago Press.

Pickering, A. (1995). *The mangle of practice: Time, agency, and science.* Chicago: University of Chicago Press.

Popper, K. R. (1959). *The logic of scientific discovery.* London: Hutchinson.

Popper, K. R. (1963). *Conjectures and refutations: The growth of scientific knowledge.* London: Routledge and Paul.

Pretty, J. N., & Chambers, R. (1993). *Towards a learning paradigm: New professionalism and institutions for agriculture.* University of Sussex Institute of Development Studies discussion paper 334. Brighton: Institute of Development studies.

Quinn, J. B., Baruch, J. J., & Zien, K. A. (1997). *Innovation explosion: Using intellect and software to revolutionize growth strategies.* New York: The Free Press.

Rahman, M. A. (1993). *People's self development: Perspectives on participatory action research; a journey through experience.* London: Zed Books, University Press.

Ravetz, J. (1999). What is post-normal science? *Futures 31*(7), 647–653.

Resource Assessment Commission. (1993). *Resource Assessment Commission coastal zone inquiry - final report.* Canberra: King's Cross NSW.

Rossiter, M. W. (1975). The emergence of agricultural science: Justus Liebig and the Americans, 1840–1880. In *Yale studies in the history of science and medicine* (Vol. 9). New Haven: Yale University Press.

Sardar, Z. (2000). Thomas Kuhn and the science wars. In R. Appinganesi (Ed.), *Postmodern encounters series.* London: Icon Books.

Scoones, I., & Thompson, J. (1994). Knowledge, power and agriculture—towards a theoretical understanding. In I. Thompson & A. J. Scoones (Eds.), *Beyond farmer first: Rural people's knowledge, agricultural research and extension practice* (pp. 16–32). London: Intermediate Technology Publications.

Slatyer, R. (2000). CRCs—a retrospective view. *Focus (Australian Academy of Sciences and Engineering), 113*(July–August), 9–14.

Sless, D. (1986). In search of semiotics. Totowa, NJ: Barnes & Noble Books.

Sokal, A. (1996). Transgressing the boundaries: Towards a transformative hermeneutics of quantum gravity. *Social Text, 46–47*, 217–252.

Stewart, J. (Ed.). (1996). Beyond the symbol model: Reflections on the representational nature of language. In D. Cahn (Ed.), *SUNY series in speech communication.* Albany: SUNY Press.

Strauss, A., & Corbin, J. (1998). *Basics of qualitative research: Techniques and procedures for developing grounded theory.* Thousand Oaks, London, New Delhi: Sage Publications.

Taylor, J. C. (1971). *Technology and planned organizational change.* Ann Arbor: Center for Research and Utilization of Scientific Knowledge, University of Michigan.

Tripp, R. (Ed.). (1992). *Planned change in farming systems: Progress in on-farm research.* John Wiley & Sons.

[UNESCO] United Nations Educational, Scientific and Cultural Organization. (1999). *The Science Agenda–a framework for action. Science for the 21st Century: A new commitment.* World Conference on Science. Budapest: UNESCO.

[UNFCCC] United Nations Framework Convention on Climate Change. (n.d.). *Kyoto protocol to the United Nations framework convention on climate change.* Retrieved April 21, 2001, from http://unfccc.int/resource/conv/

[UNCED] United Nations Commission on Environment and Development. (1992). *Agenda 21.* Online source retrieved March 30, 2001, from http://www.un.org/esa/sustdev/agenda21text.htm

Vanclay, F., & Lawrence, G. (1995). *The environmental imperative: Eco-social concerns for Australian agriculture.* Rockhampton, Queensland: Central Queensland University Press.

van Kerkhoff, L. (2005a). Integrated research: Concepts of connection in environmental science and policy. *Environmental Science & Policy, 8*(5), 439–463.

van Kerkhoff, L. (2005b). Strategic integration: The practical politics of integrated research in context. *Journal of Research Practice, 1*(2), Article M5.

Watson, R. T., Noble, I. R., Bolin, B., Ravindranath, N. H., Verardo, D. J., & Dokken, D. J. (2000). *Land use, land-use change, and forestry 2000: Special report of the Intergovernmental Panel on Climate Change.* Cambridge, UK: Intergovernmental Panel on Climate Change, Cambridge University Press.

[WCED] World Commission on Environment and Development. (1987/1990). *Our common future.* (1990 Australian ed.) Melbourne.

Weinberg, S. (2001). *Facing up: Science and its cultural adversaries.* Boston: Harvard University Press.

Wenger, E. (1998). Communities of practice: Learning, meaning and identity. In R. Pea, J. S. Brown & J. Hawkins (Eds.), *Learning in doing: Social, cognitive and computational perspectives series.* Cambridge, UK: Cambridge University Press.

Ziman, J. M. (2000). *Real science: What it is, and what it means.* Cambridge, UK: Cambridge University Press.

Printed in the United States
by Baker & Taylor Publisher Services